Hands-on JavaScript for Python Developers

Leverage your Python knowledge to quickly learn JavaScript and advance your web development career

Sonyl Nagale

BIRMINGHAM - MUMBAI

Hands-on JavaScript for Python Developers

Copyright © 2020 Packt Publishing

Commissioning Editor: Pavan Ramchandani
Acquisition Editor: Ashitosh Gupta
Content Development Editor: Divya Vijayan
Senior Editor: Keagan Carneiro
Technical Editor: Shubham Sharma
Copy Editor: Safis Editing
Project Coordinator: Kinjal Bari
Proofreader: Safis Editing
Indexer: Rekha Nair
Production Designer: Shankar Kalbhor

First published: September 2020

Production reference: 1250920

Published by Packt Publishing Ltd.
Livery Place
35 Livery Street
Birmingham
B3 2PB, UK.

ISBN 978-1-83864-812-1

www.packt.com

To those who dream.

To those who aspire.

To those who make a difference.

To those who are here now, have always been, and will always be.

To family, chosen or otherwise.

To BS, MF, and C+C.

To you.

`Packt.com`

Subscribe to our online digital library for full access to over 7,000 books and videos, as well as industry leading tools to help you plan your personal development and advance your career. For more information, please visit our website.

Why subscribe?

- Spend less time learning and more time coding with practical eBooks and Videos from over 4,000 industry professionals

- Improve your learning with Skill Plans built especially for you

- Get a free eBook or video every month

- Fully searchable for easy access to vital information

- Copy and paste, print, and bookmark content

Did you know that Packt offers eBook versions of every book published, with PDF and ePub files available? You can upgrade to the eBook version at `www.packt.com` and as a print book customer, you are entitled to a discount on the eBook copy. Get in touch with us at `customercare@packtpub.com` for more details.

At `www.packt.com`, you can also read a collection of free technical articles, sign up for a range of free newsletters, and receive exclusive discounts and offers on Packt books and eBooks.

Contributors

About the author

Chicago-born, Iowa-raised, Los Angeles-seasoned, and now New York City-flavored, **Sonyl Nagale** started his career as a graphic designer focusing on web, which led down the slippery slope to becoming a full-stack technologist instead. With an eye toward the client use case and conversation with the creative side, he prides himself on taking a holistic approach to software engineering. Having worked at start-ups and global media companies using a variety of languages and frameworks, he likes solving new and novel challenges. Passionate about education, he's always excited to have great teachable moments complete with laughter and seeing the "Aha!" moments in students' eyes.

> *I want to thank all the people who have believed in me and watched me grow. My friendships with all of you mean more than you know.*

About the reviewer

Emmanuel Demey works with the JavaScript ecosystem on a daily basis. He spends his time sharing his knowledge with anyone and everyone. His first goal at work is to help the people he works with. He has spoken at numerous French conferences (including DevFest Nantes, DevFest Toulouse, Sunny Tech, and Devoxx France) about topics related to the web platform, such as JavaScript frameworks (Angular, React.js, and Vue.js), accessibility, and Nest.js. He has been a trainer for 10 years at Worldline and Zenika (two French consulting companies). He is also the co-leader of the Google Developer Group de Lille and the co-organizer of the DevFest Lille conference.

Gil Goncalves is a software engineer who has worked with multiple programming languages, but the longest affair has been with Python. He has worked in a variety of companies from multiple fields in his 11 years in the tech industry. He's given talks at many conferences about code, tools, and engineering careers. He plays video games, dabbles in various hobbies, makes a mean sourdough bread, writes code, and organizes and attends meetups and conferences about programming.

Packt is searching for authors like you

If you're interested in becoming an author for Packt, please visit `authors.packtpub.com` and apply today. We have worked with thousands of developers and tech professionals, just like you, to help them share their insight with the global tech community. You can make a general application, apply for a specific hot topic that we are recruiting an author for, or submit your own idea.

Table of Contents

Section 3 - The Back-End: Node.js vs. Python

Preface

In learning Python, you took the first step in your software engineering career by learning Python's fundamentals, its elegance, and its programming principles. For the next stage of your career, let's learn how to transfer your programming knowledge to JavaScript to work on frontend tasks, including UX/UI work, form validation, frontend animations, and more. You may be familiar with rendering the frontend with Flask, but JavaScript will allow you to create user interfaces and react in real time to user input.

We'll go through the differences between the two languages, not just on a syntactical level, but also semantical: *why* and *when* we use JavaScript over Python, what their separations of concerns are, how we connect our existing HTML and CSS with JavaScript on the frontends and backends using Node.js to create engaging user experiences, and how we create full-stack applications utilizing all layers of a web application.

Who this book is for

One size does *not* fit all in software engineering. Python is an approachable, scalable language designed for backend web work but can open the door to curiosity about the frontend. This book is written for programmers with 1-3 years of Python experience who wish to expand their knowledge of the web ecosystem to the frontend programming world enabled by JavaScript and understand how using JavaScript on both the frontend and the backend, through Node.js, leads to efficient coding and workflows.

A solid understanding of data types, functions, and scope will be important to grasping the concepts laid out in this book. Familiarity with HTML, CSS, the **Document Object Model** (**DOM**), and Flask and/or Django will come in handy.

What this book covers

Chapter 1, *The Entrance of JavaScript into Mainstream Programming*, is where we will learn about the importance of JavaScript.

Chapter 2, *Can We Use JavaScript Server-Side? Sure!*, delves into server-side JavaScript. JavaScript's use is more than browser-side and can be used for rich, complex, server-based applications.

Chapter 3, *Nitty-Gritty Grammar*, is where you will learn the details of how to write JavaScript and how its grammar differs from Python.

Chapter 4, *Data and Your Friend, JSON*, covers data. Every computer program must work with some sort of data. You will be learning how to interact with data within JavaScript.

Chapter 5, *Hello World! and Beyond: Your First Application*, sees you write your first JavaScript programs!

Chapter 6, *The Document Object Model (DOM)*, teaches you about the basics of how to work with a web page in order to connect JavaScript with user interactions.

Chapter 7, *Events, Event-Driven Design, and APIs*, takes you beyond basic interactions and shows you how to incorporate dynamic data into your programs.

Chapter 8, *Working with Frameworks and Libraries*, introduces some of the modern scaffolds for JavaScript programs in order to expand your knowledge of industry-standard applications.

Chapter 9, *Deciphering Error Messages and Performance Leaks*, covers errors. Errors happen! We should know a bit about how to work with them and debug our programs.

Chapter 10, *JavaScript, Ruler of the Frontend*, takes a closer look at how JavaScript brings the frontend together.

Chapter 11, *What is Node.js?*, goes into Node.js. Since the use of JavaScript on the frontend has been examined, this chapter dives into its role in the "JavaScript everywhere" paradigm using Node.js.

Chapter 12, *Node.js versus Python*, asks, Why would a developer choose Node.js over Python? Can they work together? And how do we install packages that we need to create and run our programs?

Chapter 13, *Using Express*, looks at Express.js (or just Express), which is a web application framework, considered the de facto web server of Node.js.

Chapter 14, *React with Django*, explores Django. You may have Django as a Python framework, so let's see how it differs from JavaScript frameworks on both the frontend and the backend.

Chapter 15, *Combining Node.js with the Frontend*, wires together both the frontend and the backend. We'll build two small applications for (almost) full-stack functionality.

Chapter 16, *Enter Webpack*, concerns tooling for deployment, which is vital for efficient JavaScript.

Chapter 17, *Security and Keys*, dives into security. JavaScript needs knowledge of secure resources, so how do we deal with them?

Chapter 18, *Node.js and MongoDB*, moves on to MongoDB. MongoDB is a great example of how to work with a database from JavaScript. We'll be using it as our example NoSQL database, as it works well with JSON data.

Chapter 19, *Putting It All Together*, has you create a final project using a full, modern JavaScript stack.

To get the most out of this book

Since we'll be working first-hand with JavaScript, you will need to have a code editor installed on your computer, such as Visual Studio Code, Sublime Text, or another general-purpose programming environment. A mobile device such as a tablet will likely not be an appropriate environment due to the coding environment limitations, though a lower-end machine will work. We will be working with command-line tools, so familiarity with the macOS Terminal will be useful; Windows OS users should download and install Git Bash or a similar terminal program as the standard Windows Command Prompt will not be sufficient.

A modern browser will be required to work with our programs. Chrome is recommended. We will be using ECMAScript 2015 (also known as ES6) throughout our JavaScript work.

We will be installing various other components of our systems, such as Node.js and Node Package Manager, Angular, and React. Instructions for installing each requisite component will be provided in-chapter. It will likely be required to have administrator access to your system in order to complete all the installation steps.

If you are using the digital version of this book, we advise you to type the code yourself or access the code via the GitHub repository (link available in the next section). Doing so will help you avoid any potential errors related to the copying and pasting of code.

Several of our projects will also require accessing websites, so an active internet connection will be necessary. A bit of a sense of humor is also recommended.

Download the example code files

You can download the example code files for this book from your account at `www.packt.com`. If you purchased this book elsewhere, you can visit `www.packtpub.com/support` and register to have the files emailed directly to you.

You can download the code files by following these steps:

1. Log in or register at `www.packt.com`.
2. Select the **Support** tab.
3. Click on **Code Downloads**.
4. Enter the name of the book in the **Search** box and follow the onscreen instructions.

Once the file is downloaded, please make sure that you unzip or extract the folder using the latest version of:

- WinRAR/7-Zip for Windows
- Zipeg/iZip/UnRarX for Mac
- 7-Zip/PeaZip for Linux

The code bundle for the book is also hosted on GitHub at `https://github.com/PacktPublishing/Hands-on-JavaScript-for-Python-Developers`. In case there's an update to the code, it will be updated on the existing GitHub repository.

We also have other code bundles from our rich catalog of books and videos available at `https://github.com/PacktPublishing/`. Check them out!

Download the color images

We also provide a PDF file that has color images of the screenshots/diagrams used in this book. You can download it here: `https://static.packt-cdn.com/downloads/9781838648121_ColorImages.pdf`.

Conventions used

There are a number of text conventions used throughout this book.

`CodeInText`: Indicates code words in text, database table names, folder names, filenames, file extensions, pathnames, dummy URLs, user input, and Twitter handles. Here is an example: "Mount the downloaded `WebStorm-10*.dmg` disk image file as another disk in your system."

A block of code is set as follows:

```
html, body, #map {
  height: 100%;
  margin: 0;
  padding: 0
}
```

When we wish to draw your attention to a particular part of a code block, the relevant lines or items are set in bold:

```
[default]
exten => s,1,Dial(Zap/1|30)
exten => s,2,Voicemail(u100)
exten => s,102,Voicemail(b100)
exten => i,1,Voicemail(s0)
```

Any command-line input or output is written as follows:

```
$ mkdir css
$ cd css
```

Bold: Indicates a new term, an important word, or words that you see onscreen. For example, words in menus or dialog boxes appear in the text like this. Here is an example: "Select **System info** from the **Administration** panel."

Warnings or important notes appear like this.

Tips and tricks appear like this.

Get in touch

Feedback from our readers is always welcome.

General feedback: If you have questions about any aspect of this book, mention the book title in the subject of your message and email us at customercare@packtpub.com.

Errata: Although we have taken every care to ensure the accuracy of our content, mistakes do happen. If you have found a mistake in this book, we would be grateful if you would report this to us. Please visit www.packtpub.com/support/errata, selecting your book, clicking on the Errata Submission Form link, and entering the details.

Piracy: If you come across any illegal copies of our works in any form on the Internet, we would be grateful if you would provide us with the location address or website name. Please contact us at copyright@packt.com with a link to the material.

If you are interested in becoming an author: If there is a topic that you have expertise in and you are interested in either writing or contributing to a book, please visit authors.packtpub.com.

Reviews

Please leave a review. Once you have read and used this book, why not leave a review on the site that you purchased it from? Potential readers can then see and use your unbiased opinion to make purchase decisions, we at Packt can understand what you think about our products, and our authors can see your feedback on their book. Thank you!

For more information about Packt, please visit packt.com.

Section 1 - What is JavaScript? What is it not?

Ah, JavaScript, the mysterious beast. Let's unpack what it is and isn't, because the front-end can barely exist without it, and the back-end loves it.

In this section, we will cover the following chapters:

- Chapter 1, *The Entrance of JavaScript into Mainstream Programming*
- Chapter 2, *Can We Use JavaScript Server-Side? Sure!*
- Chapter 3, *Nitty-Gritty Grammar*
- Chapter 4, *Data and Your Friend, JSON*

1
The Entrance of JavaScript into Mainstream Programming

JavaScript can run both client- and server-side, which inherently means that the use cases for using JavaScript versus Python will vary. From humble beginnings, JavaScript, with its quirks, strengths, and limitations, is now one of the main pillars of the interactive web as we know it, from powering rich frontend interactions, to web servers. How did it become one of the most important ubiquitous technologies of the web? In order to grasp JavaScript's powerful ability to add functionality to both the front- and backends, we first need an understanding of what the frontend is—and what it isn't. Having an understanding of JavaScript's origins helps to clarify the "why" of JavaScript, so let's take a look.

The following topics will be covered in this chapter:

- The **National Center for Supercomputing Applications** (**NCSA**) and the need for interactivity
- Early web browsers and a 10-day prototype
- Enter Ecma International
- HTML, CSS, and JavaScript—the best friends of the frontend
- How JavaScript fits into the frontend ecosystem

Technical requirements

You can find the code files present in this chapter on GitHub at `https://github.com/PacktPublishing/Hands-on-JavaScript-for-Python-Developers`.

NCSA and the need for interactivity

The early internet was a fairly boring place compared with the rich medium we now have in the 21st century. Without graphical browsers and only fairly rudimentary (and esoteric) commands, early adopters were able to do only certain academic tasks for a period of time. Starting from **ARPANET** (the **Advanced Research Projects Agency Network**), it was designed to facilitate basic communication and file transfers by being one of the first packet-switching networks. Additionally, it was the first network to implement the **Transmission Control Protocol/Internet Protocol** (**TCP/IP**) suite, which we now take for granted as it runs behind the scenes of all modern web applications.

Why is this significant? The early internet was designed for fundamental and simple purposes, but it has grown since then. As a Python developer, you already understand the power of the modern web, so a full history of the web isn't needed. Let's skip to the origins of what we now know as the frontend.

Enter Tim Berners-Lee in 1990: the invention of the World Wide Web. By building the first web browser himself and with the **European Organization for Nuclear Research** (known as **CERN**) creating the first website, the floodgates opened and the world was never the same. What started as academic tinkering has now become a global necessity, with millions of people around the globe relying on the web. It goes without saying that today, in the 21st century, we use the web and multiple forms of digital communication to go about our everyday lives.

One of the projects that Berners-Lee created was **HTML—Hypertext Markup Language**. As the backbone of a website, this basic markup language spawned significant growth and development in the computing community. It only took a few years (the year was 1993, to be precise) for Mosaic, the first iteration of what we now call a browser, to be released. It was developed by the NCSA at the University of Illinois at Urbana-Champaign and was a vital part of the web's development.

Early web browsers and a 10-day prototype

So, why JavaScript? Obviously, the web needed more than just static data to be useful, so, in 1995, Brendan Eich at Netscape Communications came along. Originally, the idea wasn't to create a whole new language but rather to incorporate Scheme into Netscape. That idea was superseded by the work Sun Microsystems did with Java. It was decided that this language that Eich was creating would be somewhat Java-like, and not Scheme at all. The genesis of the idea came from Marc Andreessen, the founder of Netscape Communications. He felt there needed to be a language to combine HTML with a "glue language" that helped with images, plugins, and—yes—interactivity.

Eich created a prototype of JavaScript (originally called Mocha, and then LiveScript) in 10 days. It's hard to believe that a 10-day prototype has become such a crucial part of the web, but those are the facts as history records them. Once Netscape developed a production-ready version, JavaScript shipped with Netscape Navigator in 1995. Shortly after JavaScript was released, Microsoft created its own version of JavaScript, called (unceremoniously) JScript. JScript shipped with Microsoft's Internet Explorer 3.0 in 1996.

Now, there were two technologies competing for the same space. JScript was reverse-engineered from Netscape's JavaScript, but since the two flavors of the language had their own quirks, the browser wars began, leading sites to often have a label "Best viewed in Netscape Navigator" or "Best viewed in Internet Explorer", due to the technological complexities involved in supporting both technologies on one site. A portent of things to come, the differences in the early versions only increased. Some websites would work flawlessly in one browser and break horrifically in the other—not to mention the complications caused by other competitors to both Netscape's and Microsoft's browsers! Early developers also found the differences between the two technologies only fueled the arms race. If you experienced such degradation of performance (or, even worse, you were working with JavaScript in the early days, like me), you definitely felt the pain of the competing versions. Each company, as well as other third parties, raced to create the next best JavaScript release. At its core, JavaScript has to be interpreted client-side, and the differences between the browsers led to bedlam. Something had to be done, and Netscape had a solution, though it wouldn't be perfect.

We'll learn about this solution in the next section.

Enter Ecma International

The **European Computer Manufacturers Association** (**ECMA**) changed its name in 1994, becoming Ecma International, to reflect its refined purpose. As a standards organization, its purpose is to facilitate modernization and consistency for various technologies. Partly in response to Microsoft's work, Netscape approached Ecma International in 1996 to standardize the language.

JavaScript became documented in the ECMA-262 specification. You may have seen the term **ECMAScript** or "ECMAScript-based languages". There are more ECMAScript languages than just JavaScript! ActionScript is another ECMAScript-based language that follows similar conventions to JavaScript. With the downfall of Flash as a web technology, we don't see ActionScript much in practice anymore save for a few discrete uses, but the fact remains: Ecma International created standards and they have been used to create different technologies, which helped ease the browser wars—for a time.

Perhaps the most interesting part of Ecma International with regard to JavaScript is the various versions that have been codified. To date, there are nine versions, all with varying differences. We will be using ECMAScript 2015 (also known as ES6) throughout this book, as it is the most stable baseline for web development work today. Features of the 2016-2018 versions can be used by some browsers and will be introduced.

HTML, CSS, and JavaScript – the best friends of the frontend

Powering every modern website or web application are, at a minimum, three technologies: HTML, **Cascading Style Sheets** (**CSS**), and JavaScript. They are the "best friends" of the frontend, and are illustrated in the following screenshot:

Figure 1.1 - The best friends: HTML, CSS, and JavaScript

At the intersection of the three technologies is where our modern website lives. Let's take a look at these in the following sections.

HTML, the overlooked hero

When we think about the web, the basic structure of a site—the skeleton, if you will—is HTML. However, with its (purposeful) simplicity, it's often ignored as being a simple technology. One way to think about a website is thinking about a body: HTML is the skeleton; CSS is the skin; our friend JavaScript is the muscle.

HTML's history is inextricably tied to that of the web itself, as it continues to evolve with advancing specifications, features, and syntax as the web itself grows. But what is HTML? It's not a full-fledged programming language: it can't do logic or manipulate data. However, as a markup language, it's incredibly important to our use of the web. We won't spend too much time talking about HTML, but some basics will get us on the right track.

The HTML specification is controlled by the **World Wide Web Consortium (W3C)**, and its current version is HTML5. HTML's grammar consists of elements, called tags, that have specific definitions and are surrounded by angle brackets. When used in JavaScript, these tags describe nodes of data that JavaScript can read and manipulate.

Why is HTML important to us in JavaScript? JavaScript can touch HTML using the browser's internal **Application Programming Interface (API)** called the **Document Object Model (DOM)**. The DOM is the programmatic representation of all the HTML on the page, and it dictates how JavaScript can manipulate elements on a rendered page. Unlike Python, JavaScript can react to user inputs without communicating back to the server; its execution logic can happen on the frontend. Think about when you enter information in a form on a website. Sometimes, there are required fields, and if you attempt to submit the form, JavaScript can halt the submission to the server and give visual cues—such as red outlines on required boxes and a warning message—and convey to the user that information is missing. This is an example of JavaScript using the DOM for interactivity. We'll dive further into this later on, in Chapter 7, *Events, Event-Driven Design, and APIs*.

Here's an example of a simple HTML5 boilerplate:

```
<!doctype html>

<html lang="en">
<head>
  <meta charset="utf-8">

  <title>My Page</title>

</head>

<body>
  <h1>Welcome to my page!</h1>
  <p>Here's where you can learn all about me</p>
</body>
</html>
```

It's pretty legible in and of itself: contained within tags titled title is a string containing a simple title for this page. In the meta tag, we have one more element besides the name of the tag: the charset *attribute*. HTML5 also introduced *semantic* tags, which not only provide a visual structure to the page but also describe the purpose of the tag. For example, nav and footer are used to denote navigation and footer sections on a page. If you'd like to experiment with HTML, CSS, and JavaScript as we progress, you can use a tool such as Codepen.io or JSFiddle.net. Since we're so far only working with client-side work, you don't need a compiler or any other software on your computer. You can also work locally with your favorite text editor and then load your HTML in a browser.

One more set of attributes that are important to our needs with JavaScript are `class` and `id`. These attributes provide an efficient conduit by which JavaScript can access HTML. Let's take a look in the following code block at a more fleshed-out example of HTML:

```html
<!doctype html>

<html lang="en">
<head>
  <meta charset="utf-8">

  <title>My Page</title>

</head>

<body>
  <h1 id="header">Welcome to my page!</h1>
  <label for="name">Please enter your name:</label>
  <form>
    <input type="text" placeholder="Name here" name="name" id="name" />
    <p class="error hidden" id="error">Please enter your name.</p>
    <button type="submit" id="submit">Submit</button>
  </form>
</body>
</html>
```

The output of this will give us a very simple page, as follows:

Figure 1.2 - A simple HTML page

Very basic, right? Why is **Please enter your name** repeated? If you notice the second p tag on the page, one of its classes is `hidden`. However, we can still see it. We'll need CSS to help us out here.

CSS

If HTML is the bone structure of our page, then CSS is the *skin* of it, giving it a look and feel. Working with JavaScript on the frontend inherently takes into consideration CSS as well. In the example of our website form, the red outlines and warning messages are often triggered by toggling CSS classes. Here's a short example of CSS:

```
.error {
  color: red;
  font-weight: bold;
}
```

In this example, we have one CSS declaration (the `error` class, denoted as a class by the period preceding its name), and two CSS rules inside the curly braces for font color and font weight. It won't be important for now to be fully versed in CSS structure and rules, but as a JavaScript developer for the frontend, you will likely interact with CSS. For example, toggling our `error` class to make the text in our form red and bold is one way that JavaScript can trigger a message to the user, informing them that there's a problem with the form submission.

Let's add the preceding CSS into our previous HTML work. We can see this results in the following change:

Figure 1.3 - Adding a bit of CSS

Now, we can see that the rules of red and bold are being reflected, but we can still see the paragraph. Our next two CSS rules are the following ones:

```
.hidden {
  display: none;
}

.show {
  display: block;
}
```

This is a little closer to what we expect to see. But why make a paragraph just to hide it with CSS?

JavaScript

Enter our friend, JavaScript. If JavaScript is going to be the muscles of the body, it's then responsible for manipulating the bones (HTML) and the skin (CSS). Our human muscles can't do all that much to change our physical appearance, but they can certainly put us in different positions, expanding and contracting our elastic skin and manipulating the positions of our bones. With JavaScript, it's possible to rearrange content on a page, change colors, create animations, and much more. We'll be diving deeply into how JavaScript interacts with HTML and CSS because, after all, JavaScript is why we're here now, reading this book!

One of the most notable points to make about JavaScript versus Python is that, in order to make changes to a page, a Pythonic program would have to respond to input from the client side from the server, and then the browser would re-render the HTML. JavaScript avoids this by executing in the browser.

For example, in our page shown previously, if the user tries to submit the form without entering a name, JavaScript can remove the `hidden` class and add the `show` class, at which point the error message shows. This is a very simple example, but it underscores the idea that JavaScript can execute changes in the browser *without* calling back to the server. Let's put the pieces together.

The HTML is shown in the following example:

```html
<!doctype html>

<html lang="en">
<head>
  <meta charset="utf-8">

  <title>My Page</title>

</head>

<body>
  <h1 id="header">Welcome to my page!</h1>
  <form>
    <label for="name">Please enter your name:</label>
    <input type="text" placeholder="Name here" name="name" id="name" />
    <p class="error hidden" id="error">Please enter your name.</p>
    <button type="submit" id="submit">Submit</button>
```

```
  </form>
  </body>
  </html>
```

The CSS is shown in the following example:

```css
.error {
  color: red;
  font-weight: bold;
}

.hidden {
  display: none;
}

.show {
  display: block;
}
```

Now, let's write some JavaScript. This likely won't make sense yet, but if you're working along in an editor such as JSFiddle, try to place the following JavaScript in the **JS** pane and hit **Run**:

```js
document.getElementById('submit').onclick = e => {
  e.preventDefault()
  if (document.getElementById('name').value === '') {
    document.getElementById('error').classList.toggle('hidden')
    document.getElementById('error').classList.toggle('show')
  }
}
```

Now, if you run this and click **Submit** without entering any data into the box, our error message will display. Very simple so far, but congratulations! You just wrote some JavaScript! Now, how would we do this with Python? We'd have to submit the form to our backend, evaluate the inputs provided, and re-render the page with our error message.

Instead, welcome to working with the *frontend*.

How JavaScript fits into the frontend ecosystem

As you can imagine, there's more to JavaScript than simply hiding and showing elements. A powerful application is more than just a collection of script tags—JavaScript fits into an overall lifecycle and ecosystem, creating rich user experiences. We'll be diving into **single-page applications (SPAs)** using React in a Chapter 8, *Working with Frameworks and Libraries*, so, for now, let's set the stage.

If you're not familiar with the term SPA don't worry—you've already used at least a few without realizing that that's what they were. Perhaps you use Google's Gmail service. If you do, poke around at it a little bit and notice that the page doesn't appear to do hard refreshes to get information from the server. Instead, it communicates asynchronously with the server and dynamically renders content. There may be periods of waiting for content to load from the server that is often filled with a little spinning icon. The name for this underlying paradigm of loading content asynchronously from a server and sending data back is called **Ajax**.

Ajax, short for **Asynchronous JavaScript and XML**, is simply a collection of technologies and techniques used on the client side to streamline the user experience by allowing data to be fetched and sent in the background. We'll be discussing calling APIs from the frontend with Ajax a bit later, but for now, let's try a small example.

Our first Ajax application

First, we'll create a very simple Python script, using Flask. If you're not familiar with Flask yet, don't worry—we won't get into it in detail here.

Here's an example of an app.py script:

```
from flask import Flask
import os

app = Flask(__name__, static_folder=os.getcwd())


@app.route('/')
def root():
```

```
    return app.send_static_file('index.html')


@app.route('/data')
def query():
    return 'Todo...'
```

Here's our HTML with JavaScript (index.html):

```
<!doctype html>

<html lang="en">
<head>
  <meta charset="utf-8">

  <title>My Page</title>

</head>

<body>
 <h1 id="header">Welcome to my page!</h1>
 <form>
   <label for="name">Please enter your name:</label>
   <input type="text" placeholder="Name here" name="name" id="name" />
   <button type="submit" id="submit">Submit</button>
 </form>
 <script>
   document.getElementById('submit').onclick = event => {
     event.preventDefault()
     fetch('/data')
       .then(res => res.text())
       .then(response => alert(response))
       .catch(err => console.error(err))
   }
 </script>
</body>
</html>
```

Before we break this down, let's try to run it, by executing the following code:

```
$ pip install flask
$ export FLASK_APP=my_application
$ export FLASK_DEBUG=1
$ flask run
```

We should see the following screen:

Figure 1.4 - A basic Flask page

Let's click **Submit**, and the following screen should appear:

Figure 1.5 - Wiring Python to JavaScript!

We successfully displayed the text **Todo...** from Python in JavaScript! Let's take a quick look at how we did that.

Our base route (the / route) will serve our static `index.html` file. Great—now, we can see our HTML. But what about the second route, `/data`? It's simply going to return text. So far, it's not very different from any bare-bones Flask application.

Now, let's take a look at our JavaScript. There's one thing to note first and foremost: in our HTML file, we can wrap our JavaScript with `<script>` tags. While storing JavaScript in a separate file brought in with its own script tag (we'll get to that), it's convenient to include code directly in your HTML for small, quick, and non-production debugging purposes. There are times when you will insert code directly in your HTML file, but this does not happen often. For now, we'll break our best practices and play with the following snippet:

```
document.getElementById('submit').onclick = event => {
```

Hm. What is this cryptic line? It's the beginning of an ES6 arrow function. We'll dive into functions in more depth later, but for now, let's see what we can glean from this line, as follows:

- `document.getElementById('submit')`: By looking at our HTML, we can see there is an element with the ID attribute of `'submit'`: the button. So, first, we're looking for our button.
- `.onclick`: Here's an action verb for us. If you guessed that this function is designed to take action when a user clicks the button, you're correct.

As for the rest of the contents of the function, we can guess that we're doing something with an event—something regarding fetching data and then doing something with it. So, what's that something?

`alert(response)` is what we're doing with it! An `alert` is just one of those annoying little pop-up messages you see in your browser, and, with the data from Flask, we displayed it in one of those! Again, not quite *practical*, but hopefully you can see where we're going: the frontend doesn't exist in a vacuum—we can communicate back and forth between the client-side and the server-side with just a few lines of code on either side.

We'll take a look at the `fetch` function in closer detail when discussing APIs, but for now, let's take a minute to look at what we've done so far with this exercise, as follows:

1. We created a small web application using Python and Flask to serve a simple HTML page.
2. This application also had an endpoint to serve a very simple message as its output: **Todo...**.
3. Using JavaScript, we took action when the user clicked the **Submit** button.
4. Upon clicking the **Submit** button, JavaScript communicated with the Python application to request data.
5. The returned data was displayed to the user in an alert window.

And that's it! We made our first successful Ajax call.

JavaScript in practice

Now that we've seen a hands-on example of how JavaScript can be used with Python, let's discuss its uses across the frontend spectrum. Spoiler alert: we'll be using JavaScript on the server side, starting in the next chapter. We encountered some cryptic commands in our Ajax example, so while it may be easy to overlook the use of and need for JavaScript, we see it's a real language with real applications.

Part of the beauty of JavaScript is its nearly universal adoption in browsers. Over time, JavaScript syntax and features have slowly evolved, but support for the different features, which were at one time wildly different between browsers, is now standardizing. Some differences still remain, however, but there are useful tools on the web that are kept up to date on the various features that browsers may or may not support. One of these sites is `caniuse.com`, and this is illustrated in the following screenshot:

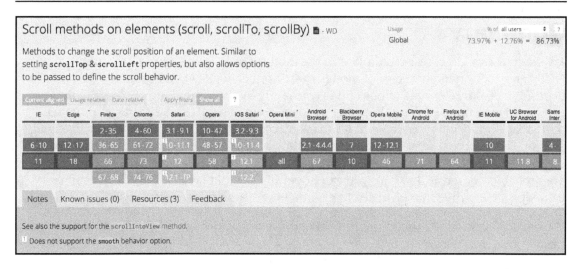

Figure 1.6: Screenshot of caniuse.com showing the selection for scroll methods on elements

The various methods and properties of JavaScript are broken out in this site by various popular browsers in a matrix to show which are (and which are not) supported by each browser. In general, however, you won't need to worry too much about whether or not your code will run on a given browser unless you're using cutting-edge features.

Now, we've shown an example of JavaScript interacting with Python as our backend using Flask, but we can use practically any backend system, as long as it's prepared to accept inbound HTTP traffic. Python, PHP, Ruby, Java—the possibilities are all there, as long as the backend expects to be working with a frontend.

A note about libraries such as jQuery: we won't be using jQuery in this book. While useful for shortcuts and simplification of some methods, one of its major draws (at least for many developers such as myself) was its under-the-hood standardization of JavaScript between browsers. Remember that Ajax `fetch` call we made? It used to be the case that an Ajax call had to be written in two different ways, one for each main type of JavaScript interpreter. However, browser standardization has eased most of the cross-browser nightmares. jQuery still provides many tools that are useful, especially for the **user interface** (**UI**), such as plugins that make it unnecessary to write components from scratch. Whether or not you use jQuery or similar libraries is up to you or will be dictated by the project's needs. Libraries such as React, which we *will* be discussing, are designed to fill a very different need than libraries such as jQuery.

Summary

JavaScript has a large, respected place in the modern web. From simple beginnings at NCSA, it's now an integral part of modern web applications, whether for UI, Ajax, or other needs. It has official specifications and is constantly evolving, making working with JavaScript ever more exciting. Working in concert with HTML and CSS, it can do much more than simple interactivity, and it can easily communicate with (almost) any backend system. Its purpose is to give us more than static pages—we want pages that do work. If you coded along, we made a simple Ajax application, and while right now the commands are probably nonsensical to you, you can hopefully see that JavaScript is fairly legible. We'll be taking a deep dive into JavaScript's grammar and construction later on.

We haven't spent time discussing the backend use of JavaScript yet, but don't worry—that's next.

Questions

Try your hand at answering the following questions to test your knowledge:

1. Which international group maintains the official specification for JavaScript?
    1. W3C
    2. Ecma International
    3. Netscape
    4. Sun

2. Which backends can communicate with JavaScript?
    1. PHP
    2. Python
    3. Java
    4. All of the above

3. Who was the original author of JavaScript?
    1. Tim Berners-Lee
    2. Brendan Eich
    3. Linus Torvalds
    4. Bill Gates

4. What is the DOM?
    1. JavaScript's representation of HTML in memory
    2. An API to allow JavaScript to modify the page
    3. Both of the above
    4. None of the above

5. What is the primary use of Ajax?
    1. Communicating with the DOM
    2. Manipulating the DOM
    3. Listening for user input
    4. Communicating with a backend

Further reading

Here are a few resources for you to peruse:

- Thoriq Firdaus, Ben Frain, and Benjamin LaGrone. *HTML5 and CSS3: Building Responsive Websites*. Birmingham: Packt Publishing, 2016.
- Browser wars: `https://en.wikipedia.org/wiki/Browser_wars`
- W3C: `https://www.w3.org/`

2
Can We Use JavaScript Server-Side? Sure!

We don't typically think of JavaScript existing server-side, as the majority of its history has only been client-side in a browser. However, at the end of the day, JavaScript *is* a language—and languages can be agnostic to their application (to an extent). While it was possible to use JavaScript server-side since its beginning with a few different tools, the introduction of **Node.js** brought using JavaScript on the server-side into the mainstream. There are more similarities between Python and JavaScript here than on the frontend, but there are still significant differences between how each technology is used in practice. Let's take a look at Node.js and how we can leverage its power on the server-side—and why we would want to!

The following topics will be covered in this chapter:

- Why use JavaScript on the server-side?
- The Node.js ecosystem
- Threading and asynchronicity

Technical requirements

You can find the code files present in this chapter on GitHub at `https://github.com/PacktPublishing/Hands-on-JavaScript-for-Python-Developers`.

Why use JavaScript on the server side?

There are many server-side languages: Java, PHP, Ruby, Go, and our friend Python, just to name a few. So, why would we want to use JavaScript as a server-side language? One answer is to reduce context switching. In theory, the same developer can write both the front- and backend of a web application with a minimum of mental changes. The research behind the cost of switching programming languages is light so far and tends to be highly anecdotal, but some studies have shown that the cognitive overhead of switching from one task to another and back again reduces productivity and increases the length of time it takes to complete a task. By extension, switching from JavaScript to Python requires a few mental gymnastics. Of course, with practice, this mental overhead becomes unimportant (think of a translator who can in real time listen to one language and translate this to a different language). However, with the speed at which technology changes, reaching that level of fluency is harder. It stands to reason that the more consistency between tasks, the less the mental overhead involved in switching between the tasks.

Let's take a look at grammatical similarities between the coding languages we've discussed in terms of syntax and style, and a bit more history.

Grammatical similarities

One of the reasons that developers enjoy working with Node.js is that it's syntactically virtually identical to frontend JavaScript.

Let's take a look at some of the code we've already written.

Here is an example of JavaScript code:

```
document.getElementById('submit').onclick = event => {
  event.preventDefault()
  fetch('/data')
    .then(res => res.text())
    .then(response => alert(response))
    .catch(err => console.error(err))
}
```

Now, let's take a look at some Node.js code that does something completely different, but with similar grammar, with dot notation, curly braces, and such. Here is an example of this:

```
const http = require('http')

http.createServer((request, response) => {
  response.writeHead(200, {'Content-Type': 'text/plain'})
```

```
    response.end('Hello World!')
}).listen(8080)
```

At first glance, these two code snippets may not look all that similar, so let's take a closer look. In our JavaScript example, take a look at `event.preventDefault()`, and then, in our Node.js example, the line `response.end('Hello World!')`. They both use **dot syntax** to specify a **method** (or function) of the parent object. These two lines are doing completely different things, but we can read them both according to the rules of JavaScript. Dot syntax is a very important concept in JavaScript, as it is inherently an object-oriented language. Much like we would when accessing class methods and properties when working with object-oriented Python, we have access to the class methods and properties of a JavaScript object. Just like in Python, we have classes, instances, methods, and properties in JavaScript.

So, what exactly *is* this Node.js example doing? Once again, we can see that JavaScript is a fairly legible language! Even without knowing too much about the innards of Node.js, we can see that we're creating a server, sending something, and listening for input. If we again compare to a Flask example, as follows, here's what we're doing:

```python
from flask import Flask, Response

app = Flask(__name__)


@app.route('/')
def main():
    content = {'Hello World!'}
    return Response(content, status=200, mimetype='text/plain')

$ flask run --port=8080
```

There's nothing inherently different about the workings of these two snippets; they are two different ways to accomplish the same goal in two different languages.

Let's take a look at a function that does the same work in client-side JavaScript and in Node.js. We haven't gone into detail on grammar quite yet, so, for the moment, don't let the syntax be a stumbling block.

Here is a JavaScript example:

```javascript
for (let i = 0; i < 100; i++) {
  console.log(i)
}
```

Here is a Node.js example:

```
for (let i = 0; i < 100; i++) {
  console.log(i)
}
```

Look closely at the two. This isn't a trick: they are, in fact, identical. Compare the JavaScript version with a basic Python loop, illustrated in the following code snippet:

```
for x in range(100):
    print(x)
```

We'll get into the grammar of JavaScript and why it appears longer than its Pythonic counterpart in Chapter 3, *Nitty-Gritty Grammar*, but for now, let's acknowledge how *different* the Python code is from JavaScript.

A bit more history

Node.js, created by Ryan Dahl and originally released in 2009, is an open source runtime for JavaScript that runs outside of a browser. It may seem new, but it has gained a large foothold in its time, including major corporations. One fact that most people don't know, however, is that Node.js is *not* the first implementation of server-side JavaScript. That distinction again belongs to Netscape, years prior. However, many considered the language not developed enough, so its usage in this vein was limited to the point of nonexistence.

Dahl sought to bring the server side and the client side closer together. Historically, there was quite a separation of concerns between the two sides of the application. JavaScript could work with the frontend, but querying the server was a continual process. The story goes that Dahl was inspired to create Node.js when he became frustrated that file upload progress bars had to rely on constant communication with the server. Node.js presents a smoother way of performing this communication by presenting an *event loop-based architecture* to facilitate this communication. Since creating Node.js, Dahl has gone on to create Deno, a JavaScript and TypeScript runtime similar to Node.js. However, for our purposes, we'll be using Node.js.

We'll get into the callback paradigm used by Node.js later, and we'll also see how frontend JavaScript uses it too.

Let's take a look at how Node.js works by taking a closer look at its proverbial life cycle.

The Node.js ecosystem

Most languages aren't of the paradigm: of just writing self-contained code. Independent modules of code, called **packages**, are widely used in software engineering and development. To think of this in another way, even a fresh web server doesn't have software on it to serve a site out of the box. You have to install a package of software, such as Apache or nginx, to even get to the "Hello World!" step of a website. Node.js is no different. It has a number of tools to make the process of getting these packages simpler, though. Let's take a look from the ground up at a basic "Hello World!" example of a server using Node.js. We'll be discussing these concepts in more detail later, so, for now, let's just go through the basic setup.

Node.js

Of course, the first thing we need is access to the language itself. There are a few methods by which you can get Node.js on your machine, including package managers, but the most straightforward way is just to download it from the official site: `https://nodejs.org`. You'll also want some familiarity with your Terminal program and basic commands. Be sure to include **Node Package Manager (npm)** when installing. Depending on your environment, you may need to reboot your machine when the installation is complete.

Once you've installed Node.js, ensure that you have access to it. Open your Terminal and execute the following command:

```
$ node -v
```

You should see a version number returned. If so, you're ready to move on!

npm

One of the powers of Node.js is its rich open source community. Of course, this isn't in any way unique to Node.js, but it is an attractive fact. Just as there is `pip` for Python, there is `npm` for Node.js. With hundreds of thousands of packages and billions of downloads, `npm` is the largest package registry in the world. Of course, with packages come a web of interdependencies and the need to keep them up to date, so npm provides a reasonably stable version management method to ensure that the packages you use function together properly in concert.

Just as we tested our Node version, we'll test npm, like this:

```
$ npm -v
```

If for some reason you do *not* have npm installed, it's time to do some research on how to install it, since the original install of Node didn't come with npm. There are several ways to install it, such as with Homebrew, but it may be best to revisit how you installed Node.

Express.js

Express is a fast, popular web application framework. We'll be using it as the basis of our Node.js work. We'll discuss using it in more detail later, so for now, let's give ourselves a quick scaffold upon which to work. We're going to install Express and a scaffolding tool globally, as follows:

1. Use the command line to install the Express generator, by running the following command: npm install -g express express-generator.
2. Use the generator to create a new directory and scaffold the application, by running the following command: express --view=hbs sample && cd sample.
3. Your sample directory should now contain a skeleton, like this:

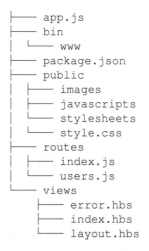

```
├── app.js
├── bin
│   └── www
├── package.json
├── public
│   ├── images
│   ├── javascripts
│   └── stylesheets
│       └── style.css
├── routes
│   ├── index.js
│   └── users.js
└── views
    ├── error.hbs
    ├── index.hbs
    └── layout.hbs
```

4. Now, we'll install the dependencies of the application by running the following command: npm install.
5. It'll do some work downloading the necessary packages, and then we'll be ready to start the server, by running the following command: npm start.

6. Visit `http://localhost:3000/` and you should see the most exciting page of all time, as shown in the following screenshot:

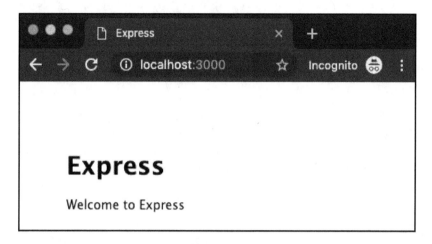

Figure 2.1 - Express welcome page

Congratulations! It's your first Node.js application! Let's take a look under the hood:

Open the `index.js` file in the `routes` directory, and you should see something similar to this:

```
var express = require('express');
var router = express.Router();

/* GET home page. */
router.get('/', function(req, res, next) {
  res.render('index', { title: 'Express' });
});

module.exports = router;
```

It's worth noting at this point that you may see a difference in syntax between some Node.js examples and modern JavaScript. If you notice, these lines end with semicolons, whereas our previous examples did not. We'll get into a discussion of the different versions of JavaScript later, but for now, just keep that note in mind if it surprised you.

Let's take a look at the `router.get` statement, illustrated in the following code block:

```
router.get('/', function(req, res, next) {
  res.render('index', { title: 'Express' });
});
```

get is referring to the HTTP verb to which the program is responding. Similarly, if we were dealing with POST data, the beginning of the line would be router.post. So, in essence, this is saying: "Hey server, when you get a request for the home page, render the index template with the title variable equal to Express." Don't worry—we'll go into much more detail on this in Chapter 13, *Using Express*, but for now, let's play around a little:

1. Add the line console.log('hello') before the res.render line.
2. Change Express to My Site.

When making changes to Node.js code, you'll need to restart the local server. You can go back to your Terminal and use *Ctrl + C* to quit Express and then npm start to restart it. Of course, there are process managers to handle this for you, but for now, we're using a very bare-bones implementation.

Navigate to https://localhost:3000/ again. You should see the following:

Figure 2.2 - Changed Express page

Now, let's go back to your Terminal. When you hit your localhost, you also triggered a `console.log()` statement—a debugging print statement. You should see `hello` in line with the requests and responses Express provided, as illustrated in the following screenshot:

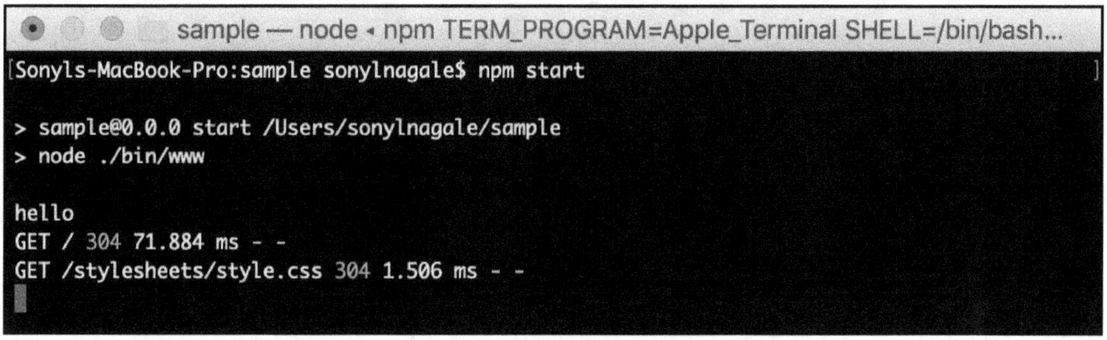

```
● ● ●          sample — node ‹ npm TERM_PROGRAM=Apple_Terminal SHELL=/bin/bash...
[Sonyls-MacBook-Pro:sample sonylnagale$ npm start

> sample@0.0.0 start /Users/sonylnagale/sample
> node ./bin/www

hello
GET / 304 71.884 ms - -
GET /stylesheets/style.css 304 1.506 ms - -
```

Figure 2.3 - console.log

Using the console will prove invaluable to us, both on the client side and the server side. This is just a taste of what it can do! Go ahead and quit with *Ctrl + C*.

Threading and asynchronicity

As with traditional web architectures, it's important to understand the *why* of using Node.js on the backend.

We've taken a look at the *how* of running Node.js, so now, let's take a look at how Node's client-server architecture differs from the traditional paradigm.

Traditional client-server architecture

To understand how Node.js differs from traditional architectures, let's look at the following request diagram:

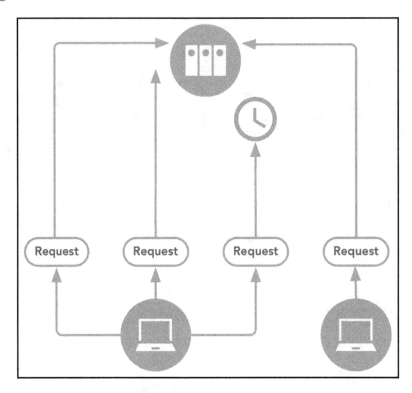

Figure 2.4 - Traditional client-server diagram

In a traditional setup, each request (or connection) to the server spawns a new thread in memory on the server, taking up system **random-access memory (RAM)** until the number of possible threads is reached. After that, some requests must wait until more memory is available. If you're not familiar with the concept of **threads**, they're basically a small sequence of commands to run on a computer. What this *multithreaded* paradigm implies is that for each new request received by the server, a new unique place is created in memory in order to handle the request.

Now, keep in mind that a *request* is not a whole web page—a page can have dozens of requests for other supplementary assets such as images. In the following screenshot, take a look at the fact that the Google home page alone has 16 requests:

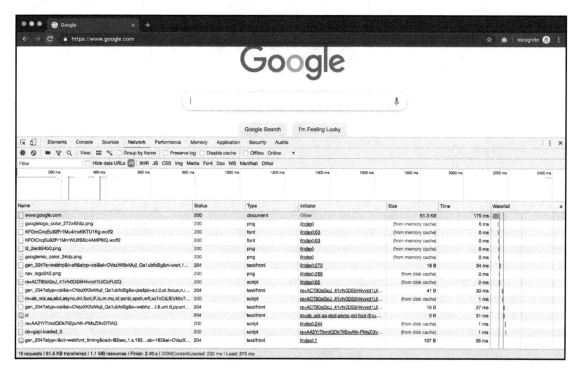

Figure 2.5 - google.com requests

Why is this important? In a nutshell: scalability. The more requests per second, the more memory is being used. We've all seen what happens when a website crashes under load—a nasty error page. This is something we all want to avoid.

Node.js architecture

In contrast to this paradigm, Node.js is *single-threaded*, allowing for thousands of non-blocking input-output calls without the need for additional overhead, as illustrated in the following diagram:

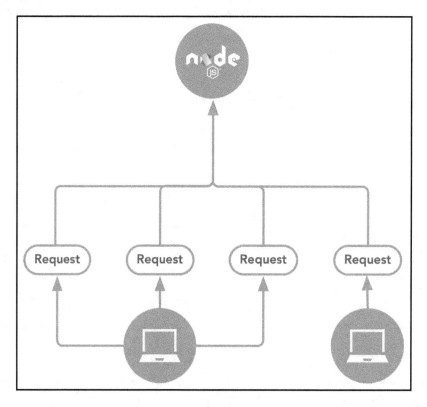

Figure 2.6 - Node.js client-server diagram

One thing to take note of early on, however: this paradigm isn't a silver bullet for managing traffic and load on a server. There really is no bullet-proof solution (yet) for the problem of large amounts of traffic. However, this structure does help a server to be more performant.

One of the reasons why Node.js pairs so well with JavaScript is that it's already dealing with the idea of **events**. As we'll see, events are a powerful cornerstone of JavaScript on the frontend, and so it stands to reason that by carrying over this process to the backend, we'll see a bit of a different approach from other architectures.

Summary

While the concept of running JavaScript on a server isn't new, its popularity, stability, and features are greatly expanded with Node.js. Early on, server-side JavaScript was abandoned but came to light again in 2009 with the creation of Node.js.

Node.js reduces the context-switching mental overhead for developers by working with the same fundamental grammar on both the client and server side. The same developer can work through the whole stack rather seamlessly because there are considerable similarities between the client-side work and how to operate with Node.js on the server. Along with a difference in approach also comes a different fundamental paradigm for handling requests to the server, compared to other more traditional implementations.

JavaScript: it's not just client-side anymore!

In the next chapter, we're going to take a deep dive into the grammar of JavaScript: syntax, semantics, and best practices.

Questions

Try your hand at answering the following questions to test your knowledge:

1. True or false: Node.js is single-threaded.
2. True or false: The architecture of Node.js makes it impervious to **Distributed Denial of Service (DDoS)** attacks.
3. Who originally created Node.js?
    1. Brendan Eich
    2. Linux Torvalds
    3. Ada Lovelace
    4. Ryan Dahl
4. True or false: JavaScript on the server side is inherently insecure because the code is exposed on the frontend.
5. True or false: Node.js is inherently superior to Python.

Further reading

Refer to the following links to get more information on this topic:

- Why The Hell Would I Use Node.js? A Case-by-Case Tutorial: `https://www.toptal.com/nodejs/why-the-hell-would-i-use-node-js`
- Event-driven architecture: `https://en.wikipedia.org/wiki/Event-driven_architecture`

3
Nitty-Gritty Grammar

When comparing two programming languages, there are bound to be structural and grammatical differences. The good news is that both Python and JavaScript are very human-readable languages, so the context switch from Python to JavaScript and Node.js shouldn't be too taxing.

Style is a good question, though: tabs or spaces? Semicolons or not? Many stylistic questions that arise when writing in any programming languages have been answered by the dicta contained within Python's PEP-8 style guide. While JavaScript doesn't have an official style guide, don't worry—it's not the Wild West out there.

Before we can write JavaScript, we must know what it is to be able to read it and understand it. All programming languages vary from one to another, and using your Python knowledge to learn a new language will require a bit of reframing of your thoughts. For example, what does JavaScript look like when we want to declare variables? How is it constructed so that the computer understands it? What do we need to watch out for as we progress?

This chapter is the key to unlocking what JavaScript can do and how to do it.

The following topics will be covered in this chapter:

- A history of style
- Grammar rules
- Punctuation and readability
- The elephant in the room – whitespace
- Existing standards – linting to the rescue!

Technical requirements

To code along with the examples in this chapter, you have a few choices:

- Code directly in the JavaScript console in your browser
- Code in the Node command line
- Use a web editor, such as `jsfiddle.net` or `codepen.io`

Using a web editor may be preferable as you can easily save your progress. You should familiarize yourself with bringing up the JavaScript console in the browser anyway, as we'll be using it for debugging output. This is usually in the **View** menu in your browser; consult your browser's documentation for how to find it if it's not immediately obvious, as some browsers require turning on **Developer** mode in **Preferences**.

You can find the code files for this chapter on GitHub at `https://github.com/PacktPublishing/Hands-on-JavaScript-for-Python-Developers/tree/master/chapter-3/Linting`.

A history of style

Every programming language has its own style, designed to ease legibility and comprehension of each line of code. Some languages are stricter than others; JavaScript in its vanilla form is one of the looser languages in adhering to style. *The Elements of Programming Style* by Brian W. Kernighan and P. J. Plauger, first published in 1974, has a number of aphorisms that have helped shape not only coding standards but also programming languages themselves.

You may be familiar with the PEP-20 aphorisms from *The Zen of Python*:

- Beautiful is better than ugly.
- Explicit is better than implicit.
- Simple is better than complex.
- Complex is better than complicated.
- Flat is better than nested.
- Sparse is better than dense.
- Readability counts.
- Special cases aren't special enough to break the rules.
- Although practicality beats purity.
- Errors should never pass silently.

- Unless explicitly silenced.
- In the face of ambiguity, refuse the temptation to guess.
- There should be one—and preferably only one—obvious way to do it.
- Although that way may not be obvious at first unless you're Dutch.
- Now is better than never.
- Although never is often better than *right* now.
- If the implementation is hard to explain, it's a bad idea.
- If the implementation is easy to explain, it may be a good idea.
- Namespaces are one honking great idea—let's do more of those!

Tongue-in-cheek qualities aside, many of these aphorisms are inspired by the principles written and experiences gained before Python was developed. First released in 1991, Python from the start had an emphasis on code readability and has some strict guidelines in place, from PEP-8 to PEP-20.

Let's take, for example, two aphorisms:

The Elements of Programming Style, 1974	The Zen of Python, 1999
Write clearly—don't be too clever.	Explicit is better than implicit.

Similar ideas are being expressed here. I think most software engineers would agree with the statement that being clear, explicit, and legible are good qualities to strive for as you develop your program.

There is one idea to keep strongly in mind, however, as you keep advancing in your JavaScript learning: since JavaScript's syntax is *designed* to be looser than that of some other languages, you may find that different companies have an in-house style to use for JavaScript code. This is not unique to JavaScript—many languages also have style guides in a company to enforce code consistency across employees. It also helps the greater ecosystem of the language to have consistently readable code. However, this does lead to differences from one code base to another in terms of style.

As with any language, we need to know our grammar to know how we're going to write JavaScript. As with Python, the machine expects properly formatted code before it does its work, and this is your job. On to grammar.

Grammar rules

Just like with any other programming language, JavaScript has grammar rules that are to be followed in order for the computer to understand what our code is trying to tell it. These rules are fairly straightforward and range from capitalizing and punctuating your code, which enhances the readability, to using specific structures within your code and avoiding common words that can confuse meaning. The rules of JavaScript syntax are fairly simple and straightforward; they include the following:

- Capitalization
- Reserved words
- Variable syntax
- Data types
- Logic structures
- Functions
- Punctuation

Capitalization counts

As with most programming languages, capitalization makes a difference. The `myNode` and `mynode` variables will be interpreted as completely different variables. That is, the computer will absolutely see the relationship between `myNode` and `mynode` because they are capitalized differently.

Reserved words

There are a good number of words reserved in JavaScript that cannot be used for variable names. Here is a list of most of them:

abstract	double	in	super
arguments	else	instanceof	switch
await	enum	int	synchronized
boolean	eval	interface	this
break	export	let	throw
byte	extends	long	throws
case	false	native	transient
catch	final	new	true
char	finally	null	try
class	float	package	typeof
const	for	private	var
continue	function	protected	void
debugger	goto	public	volatile
default	if	return	while
delete	implements	short	with
do	import	static	yield

These are always in lowercase, and if you were to attempt to use one of these words as a variable name, the program would show an error.

Declaring variables

In JavaScript, it's best practice to declare variables before using them. This declaration can happen at the time of assignment to a value, or you can define a variable without a value.

Unlike some other languages, JavaScript is *loosely typed*, so you don't need to declare what type of variable you're creating. By convention, variables in JavaScript start with a lowercase letter and follow camel-casing, rather than snake-casing. So, myAge is preferable to my_age or MyAge. Variables cannot start with a number.

There are three keywords used to declare variables in JavaScript: const, let, and var.

const

A **const**, short for **constant**, is a variable that is not expected to change in value over the course of the program. They're useful for enforcing values that you don't want to change. Prior to the sixth edition of ECMAScript, ES2015 (often called ES6), it was possible to mutate the value of any variable, so mistakes such as using an assignment operator (=) instead of a comparison operator (== or ===) were common:

```
const firstName = "Jean-Luc"
const lastName = "Picard"
```

Sure, Captain Picard *could* change his name, but that doesn't seem very likely.

Sometimes, we want to declare a variable as a hard constant, such as pi or an API key. These use cases are, in general, the only exception to the naming standards, in that they are often all uppercase and sometimes have underscores:

```
const PI = 3.14159
const API_KEY = 'cnview8773jass'
```

So far, we have examples of two data types: **strings** and **numbers**. JavaScript doesn't have a concept of *float* versus *int* versus *long*; they're all numbers. If you noticed, we can also declare strings with single or double quotes. Some libraries and frameworks have a preference for one over the other, but for standard JavaScript, it's OK to use either. It's best practice to be consistent in your usage, however.

let

When declaring a variable using `let`, we explicitly state that we expect or allow the value of the variable to change over the course of the program:

```
let ship = "Stargazer"
ship = "Enterprise" // ship now equals "Enterprise"
```

Captain Picard can be transferred to another ship at any time, so we want our program to allow changes in value.

var

The oldest way of defining a variable in JavaScript is with the `var` keyword. Declaring with `var` does not place any restrictions on the value of the variable; it can be changed.

The use of `var` is still supported but considered legacy and was deprecated in ES6. However, with decades of existing programs and examples, it's important to at least be familiar with `var`.

Data types

Even though JavaScript is loosely typed, it's important to know about the data types available to us, as we will need to know them for issues such as comparisons and reassignment.

Here is a rough mapping of the base Pythonic variables to base JavaScript:

Python	JavaScript
Number	Number
String	String
List	Array
Dictionary	Object
Set	Set

This covers the base types you're likely to use. Let's examine other, more nuanced JavaScript data types. Some have equivalents in Python and some don't:

Python	JavaScript Semi-Equivalent	Reason for Difference
bool	boolean	While the data types are identical in practice, Python's `bool` data type inherits from `int`. While `1` and `0` can be used in JavaScript to represent `True` and `False`, they will not be recognized as the `boolean` type.
None	null	Technically, `None` is an object in and of itself, whereas `null` is a falsy value.
	undefined	In JavaScript, a variable that has not been declared with a value still has a pseudovalue: the singleton of the `undefined` value.
	object	Both Python and JavaScript are object-oriented languages, but their use of objects is a bit different. The base use of an object in JavaScript is a key-value store. Objects are not primitives and can store multiple types of data.
	symbol	Symbols are a new data type in ES6. While the uses are nuanced, they're worth mentioning. They are used to create unique identifiers for objects.

Now, we need to find out a bit more about types before we can use them, including how to compare them and work with them.

typeof and equality

Even though variable types are mutable, it's often useful to know what data type a variable is at that moment. The `typeof` operator helps us do that:

```
typeof(1) // returns "number"
typeof("hello") // returns "string"
```

Notice that the return values are strings.

When comparing variables, there are two equality operators: loose and strict equality. Let's take a look at some examples:

```
let myAge = 38
const age = "38"
myAge == age
```

If we were to run this comparison, we would have the result of `true`. However, we can see that `myAge` is a number, while `age` is a string. The reason the result is `true` is that when using the loose equality operator (the double-equals), JavaScript uses *type coercion* in an attempt to be helpful. When comparing variables of different types, the values are loosely compared, so while `38` and `"38"` are different types, the result of the comparison is truthy because of their values.

As you can imagine, this can have some unexpected behavior. To ask JavaScript to include the type in the comparison, use the *strict equality* operator: the triple-equals.

With our preceding example, we can try `myAge === age` and will get the result of `false` because they are different data types. It's usually considered best practice to use strict equality to avoid type coercion unless you have a specific need to use loose equality.

Arrays and objects

Arrays and objects are not primitives and can contain mixed types. Here are a few examples:

```
const officers = ['Riker','Data','Worf']

const captain = {
  "name": "Jean-Luc Picard",
```

```
    "age": 62,
    "serialNumber": "SP 937-215",
    "command": "NCC 1701-D",
    "seniorStaff": officers
}
```

`officers` is an **array,** as we can see with the square brackets. One of the interesting facts about arrays is that even though we usually declare them as a const, the values *inside* the array can be changed. `.push()` and `.pop()` are two useful methods for manipulating arrays:

```
officers.push('Troi') // officers now equals ['Riker','Data','Worf',
    'Troi']
```

Notice that the values in the array are not ordered in any way; we can get `Riker` by accessing the array with bracket notation: `officers[0]`. However, if we were to try to completely reassign the array, we would still get an error when reassigning a declared const. Arrays can hold any combination of data types.

One very handy property of arrays that we'll be using is `.length`. Since it's a property, it does not use parentheses:

```
officers.length // now equals 4
```

Note that even though arrays are zero-indexed, the `length` property is not. There *are* four elements in the array, with *indices* from 0 to 3.

We'll discuss methods and properties more throughout this chapter.

Objects are a very strong foundational component of JavaScript. In fact, technically (almost) everything in JavaScript *is* an object! Our array methods can be accessed via dot notation specifically because an array is technically a type of object. We can't, however, access the *values* of an array with dot notation.

If we look at `captain`, we can see three different data types: string, number, and array. Objects can have nested objects as well. As part of their function as key-value stores, the key should be a string. To access a value, we use dot notation:

```
captain.command // equals "NCC 1701-D"
```

We can access parts of an object with dot notation, which is similar to a **dict** in Python, but not quite! The nuances will become more clear as we work with objects since they are fundamental to what makes JavaScript unique.

Conditionals

Let's take a look at an `if/else` statement written in two ways in both Python and JavaScript:

Python	JavaScript
	`let min`

`if a < b:`
`    min = a`
`else:`
`    min = b`
	`if (a < b) {`
	`    min = a`
	`} else {`
	`    min = b`
	`}`

`min = a if a < b else b`	`let min = (a < b) ? a : b`

In both columns, the code is doing the same thing: a simple test to see whether `a` is less than `b` and then assigning the smaller value to the `min` variable. The first row is a full `if/else` statement and the second row uses the ternary structure. There are a few grammar rules to note in these examples:

- `min` must be declared before use, as a best practice. In strict mode, this would actually throw an error.
- Our `if` clause is encapsulated with parentheses.
- Our `if/else` statements are encapsulated with curly braces.
- The keywords and operators in the ternary are significantly different (and a bit more cryptic) than in Python.

If we wanted to use what we now know about `typeof`, we can use strict equality to understand our variables a bit more:

```
let myVar = 2

if (typeof(myVar) === "number") {
  myVar++; // myVar now equals 3
}
```

Loops

There are four main types of loops in JavaScript: `for`, `while`, `do/while`, and `for..in`. (There are a couple of other ways to structure a loop, but these are the main ones.) Their use cases shouldn't be much of a surprise.

for loops

Use an iterator to execute code a specified number of times:

Python	JavaScript

`names = ["Alice","Bob","Carol"]`
`for x in names:`
`    print(x)`
	`const names = ["Alice","Bob","Carol"]`
	
	`for (let i = 0; i < names.length; i++) {`
	`    console.log(names[i])`
	`}`

Now, you may be wondering, "if JavaScript has a `for..in` loop, why aren't we using it?". As it turns out, `for/in` of Python and `for..in` of JavaScript are *false cognates*: their names look alike but are very different in use. We'll discuss JavaScript's `for..in` loop shortly. Also, note how we needed to have three clauses in our `for` loop:

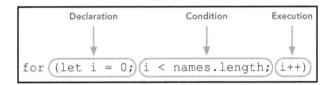

Figure 3.1 - Declaration, Condition, and Execution stages of a for loop

The **declaration** will either define an iterator or use an existing mutable variable. Note that it should be a mutable number!

Our **condition** is what we're testing. We want our loop to run while `i` is less than `names.length`. Since `name.length` is 3, we will run our loop three times, or until `i` equals 4, which no longer meets our condition.

At the end of every iteration of our loop, we **execute** something; usually a simple incrementing of our declaration. Now, notice the semicolons in between each of our clauses…unlike in other parts of JavaScript, these are *not* optional. There isn't one placed after the execution portion.

while loops

The JavaScript `while` loop is identical in use to its Pythonic equivalent, minus a little grammar:

Python	JavaScript

`i = 0`
`while i < 10:`
`    i += 1`
	`let i = 0`
	`while (i < 10) {`
	`    i++`
	`}`

do/while loops

As the name implies, the do/while loop executes the do code when a given condition equals `true`. Take a look at the JavaScript:

```
let i = 0

do {
  i++
} while (i < 10)
```

for..in loops

Now, I promised to explain why Python's `for..in` is different from JavaScript's usage. The difference is that JavaScript's `for..in` is used to iterate over keys in an object, whereas Python's `for..in` is used as a loop over a discrete set of entities.

Let's look at an example:

```
const officers = ['Riker','Data','Worf']

const captain = {
  "name": "Jean-Luc Picard",
  "age": 62,
  "serialNumber": "SP 937-215",
  "command": "NCC 1701-D",
  "seniorStaff": officers
}

let myString = ''

for (let x in captain) {
  myString += captain[x] + ' '
}
```

What do you think `myString` now equals? Since the purpose of `for..in` in *JavaScript* is to go over each *key* in the object, it's `Jean-Luc Picard 62 SP 937-215 NCC 1701-D Riker,Data,Worf`.

for..of loops

There is one more `for` loop: `for..of`, which differs from `for..in`. The `for..of` loop iterates through values of any iterable, such as arrays, strings, sets, and similar. If we want to iterate over `officers` and log out each name, we can do this:

```
for (const officer of officers) {
   console.log(officer)
}
```

Next up, we'll discuss functions!

Functions

Ah, functions. We love them because they're the key to modular, **don't-repeat-yourself (DRY)** programs. The use cases in JavaScript and Python are the same: blocks of code intended to be called more than once, usually with varying parameters. Parameters are the variables that a function will take in order to execute its code on a mutable dataset. Arguments are what we pass when we call a function. They're the same thing in essence, but have different words depending on where and when they're used: are they the abstraction, or are they the actual data? Let's take a look at a side-by-side comparison:

Python	JavaScript

`def add_one(x):`
`    x += 1`
`    return x`
 	
`print(add_one(5))`
`// output is `**`6`**
	`function addOne(val) {`
	`    return ++val`
	`}`
	
	`console.log(addOne(5))`
	`// output is `**`6`**

If you haven't already brought up the JavaScript console in your browser, you should do that now to see our output of `6`.

You can see that the structure is fairly similar, with our parameter being passed in parentheses. As noted before, we prefer camel-case in JavaScript and encapsulate with curly braces. Calling the function with our argument of 5 is the same. For conciseness, we can increment `val` with the `++` operator on the left before `return` executes. Such shortcuts are common in JavaScript, but remember to use them judiciously: "Write clearly—don't be too clever."

However, JavaScript actually has two different ways to declare a function, plus a newer syntax introduced in ES6.

Function declarations

`addOne()` in the preceding code is an example of a *function declaration*. It uses the function keyword to declare our functionality. Its anatomy is just as simple as it looks:

```
function functionName(optionalParameters, separatedByCommas) {
  // do work
  // optionally return a value
}
```

Function expressions

Here's an example of `addOne()`, constructed as a function expression:

```
const addOne = function(val) {
  return ++val
}
```

Function expressions should use `const` in the expression, though it is not syntactically incorrect to use `var` or `let`.

What's the difference between declarations and expressions? The core difference is that a function *declaration* can be used anywhere in your program because it's *hoisted* to the top. As JavaScript is interpreted top-down; this is a major exception to that paradigm. So, conversely, using an *expression* must occur after the expression is written.

Arrow functions

ES6 introduced the arrow syntax of writing function expressions:

```
const addOne = (val) => { return ++val }
```

To further complicate matters, we can omit the parentheses around `val` because there's only one parameter:

```
const addOne = val => { return ++val }
```

The main difference between arrow functions and expressions is centered around *lexical scoping*. We touched on scope with *hoisting*, and we'll discuss it in more detail in the next chapter.

Comments

As with any language, comments are important. There are two ways to declare comments in JavaScript:

```
const addOne = (val) => { return ++val } // I am an inline, single line
comment

// I am a single comment

/*
 I am a multiline comment
*/
```

So, we can start a comment with `//` and write text until the end of the line. We can have a full-line comment with `//` and we can also have a multiline comment with `/*`, ending with `*/`. Additionally, you may encounter comments in the JSDoc style, used for inline documentation:

```
/**
 * Returns the argument incremented by one
 * @example
 * // returns 6
 * addOne(5);
 * @returns {Number} Returns the value of the argument incremented by one.
 */
```

More information on JSDoc is included in the *Further reading* section.

Methods and properties

So far, we've seen `.push()` and `.pop()` as methods of array instances. In JavaScript, a **method** is simply a function inherent to its data type that operates on the data and properties of a variable. I mentioned before that nearly everything in JavaScript is an object, and that is not an exaggeration. From functionality and syntax to structure and usage, there are many similarities between the raw data type of an *object* and any other variable.

The next part of our understanding of the syntax of JavaScript is everyone's favorite: punctuation. While it may seem trivial, it's very important for the interpretation, by both humans and computers, of the code.

Punctuation and readability

As with every language, JavaScript has conventions on punctuation and how spacing affects readability. Let's take a look at a few ideas:

- **Python**:

```python
def add_one(x):
    x += 1
    return x
```

- **Java**:

```java
int add_one(int val) {
    val += 1;
    return val;
}
```

- **C++**:

```cpp
int add_one (int val)
{
    val += 1;
    return val;
}
```

- **JavaScript**:

```javascript
function addOne(val) {
    return ++val
}
```

In JavaScript, the conventions of the preceding example are as follows:

- No space between the function name and the parentheses.
- A single space before the curly brace, which is on the same line.
- The closing curly brace is on its own line, aligned with the opening statement of `function`.

There's also one more modern point to make here about JavaScript and the examples we'll be using in this book versus what you may encounter in the field and examples online: **semicolons**.

With few exceptions, in modern JavaScript, semicolons at the end of statements are *optional*. It used to be a best practice to always terminate statement lines with semicolons, and you'll see a lot of semicolons in existing code. This is a question of style from company to company, project to project, and library to library. There are some standards in place, as we'll discuss shortly with linting, but for the purposes of this book, we will *not* be using semicolons to terminate statements, except in such cases as required by syntax (such as we saw in our loops).

It's also important to note that nested lines should be indented by two spaces. Two versus four is a style question, also, but in this book, we'll be using **two spaces**. One way to help maintain consistency is to configure your code editor to translate tabs into two spaces (or four, as desired). That way, you just hit *Tab* instead of worrying about how many times you mashed the space bar. I won't expound on the importance of proper indentation, but remember: the more your code adheres to styles and best practices, the more legible it will be to those who maintain your code—and to your future self!

The elephant in the room – whitespace

OK, OK, we know that Python is whitespace-delimited: tabs matter! However, JavaScript really *doesn't* care about whitespace in most cases. As we saw before, indentation and whitespace is a matter of *style*, not *syntax*.

So here's the thing: when I was first learning Python, the idea of a language that was whitespace-dependent was abhorrent. "How could a language that could break with an improper IDE setting survive?", I thought. My opinions aside, the good news is that indentation in Python is parallel to indentation plus curly braces in JavaScript.

Here's an example:

Python	JavaScript
| ```
def hello_world(x):
 if x > 3:
   x += 1
 else:
   x += 2
 return x
``` | ```
function helloWorld(val) {
  if (val > 3) {
    return ++val
  } else {
    return val+2
  }
}
``` |

If you notice, our `if` statement inside our Python function is indented in the same way that this JavaScript example is indented, albeit without the curly braces. So yay! Your adherence to Pythonic indentation dicta actually comes *quite* in handy in JavaScript! While it's not required to include whitespace *à la* Python, it does definitely improve legibility.

At the end of the day, JavaScript likes indentation as much as Python because it makes for more readable code, though it's not required for your program to run.

Existing standards – linting to the rescue!

We've looked at JavaScript's conventions and norms, but most rules have a caveat that "this could vary" or "this isn't technically required." So, how do we make sense of our code in a malleable, opinion-driven environment? One answer: *linting*.

Simply put, **linting** refers to the process of running your code through predefined rules to ensure that not only is it syntactically correct, but it also adheres to proper style rules. This isn't a practice limited to JavaScript; you may have linted your Python code, too. In modern JavaScript, linting has come to be seen as a best practice to ensure your code is consistent. Two of the main style guides in the community are AirBnB (`https://github.com/airbnb/javascript`) and Google (`https://google.github.io/styleguide/jsguide.html`). Your code editor probably supports using a linter, but we won't go into using them in practice right now, as each editor varies in setup. Here's a quick look at what it can look like in Atom:

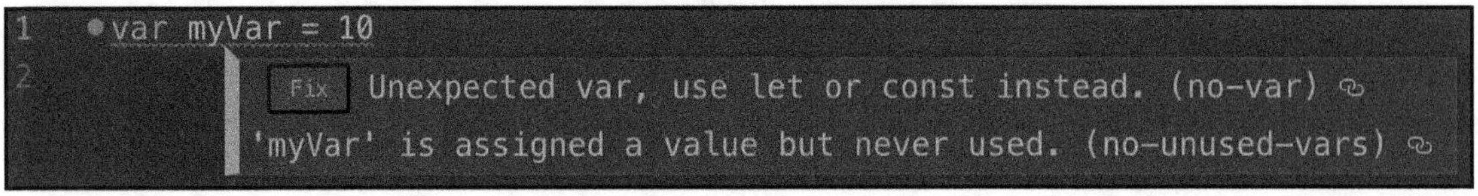

```
1  ●var myVar = 10
2          Fix  Unexpected var, use let or const instead. (no-var) ⟲
           'myVar' is assigned a value but never used. (no-unused-vars) ⟲
```

Figure 3.2 - Linting error in Atom

For our purposes, know that standards do exist, though they may vary from style guide to style guide. You can clone a repository demonstrating linting from `https://github.com/PacktPublishing/Hands-on-JavaScript-for-Python-Developers/tree/master/chapter-3/Linting`.

There are several popular linting tools available, such as ESLint and Prettier. The tool you choose can be customized to the style guide you choose to work with.

OK, that's a lot for one chapter! Let's wrap up.

Summary

JavaScript has a rich grammar and syntax, developed over years of use and refinement. Using ES6, we have a whole host of data types, methods of declaring functions, and code standards. While writing JavaScript can seem to be playing very loose and fast, there are best practices, and the fundamentals of the language are as robust as other languages. Remember that capitalization counts; don't use reserved words for variable names; use `const` or `let` to declare your variables; even though JavaScript is loosely typed, data types are important; and conditionals, loops, and functions all help structure the logic of your code.

A mastery of the grammar and syntax of JavaScript is vital to understanding how to use this robust language, so take your time familiarizing yourself with the details and intricacies. Moving forward, we will presume you have fluency in JavaScript's style as we get into more difficult material.

In the next chapter, we will get our hands dirty with data and understanding how JavaScript works with and models data.

Questions

Try your hand at answering the following questions to test your knowledge:

1. Which of the following is not a valid JavaScript variable declaration?
    1. `var myVar = 'hello';`
    2. `const myVar = "hello"`
    3. `String myVar = "hello";`
    4. `let myVar = "hello"`

2. Which of these starts a function declaration?
    1. `function`
    2. `const`
    3. `func`
    4. `def`

3. Which of these is not a basic loop type?
    1. `for..in`
    2. `for`
    3. `while`
    4. `map`

4. True or false – JavaScript *requires* line delineation with semicolons.
    1. True
    2. False

5. True or false – whitespace *never* counts in JavaScript.
    1. True
    2. False

Further reading

- B. W. Kernighan and P. J. Plauger, *The Elements of Programming Style 2nd Edition*, McGraw Hill, New York, 1978. ISBN 0-07-034207-5
- PEP-8 – *A Style Guide for Python Code*: https://www.python.org/dev/peps/pep-0008/
- PEP-20 – *The Zen of Python*: https://www.python.org/dev/peps/pep-0020/
- JSDoc: http://usejsdoc.org/

Data and Your Friend, JSON

4

It's time to learn the specifics of how JavaScript deals with data internally. Most of these structures are (nearly) identical to Python, but there are differences in syntax and usage. We touched on them in Chapter 3, *Nitty-Gritty Grammar*, but now it's time to take a deeper dive into how we work with data and use methods and properties. Understanding how to work with data is foundational to using JavaScript, especially when doing advanced work such as working with APIs and Ajax.

The following topics will be covered in this chapter:

- Data types – both JavaScript and Python are dynamically typed!
- Exploring data types
- Arrays and sets
- Objects and JSON
- The HTTP verbs
- API calls from the frontend – Ajax

Technical requirements

Clone or download the repository for this book from GitHub at https://github.com/ PacktPublishing/Hands-on-JavaScript-for-Python-Developers and look through the Chapter-4 material.

Data types – both JavaScript and Python are dynamically typed!

In Chapter 3, *Nitty-Gritty Grammar*, we discussed using typeof() to ascertain what a variable's data type is and using let and const to define them. There's an interesting fact about JavaScript that Python shares: both are dynamically typed. As opposed to statically typed languages such as Java, JavaScript's variable types can change over the course of a program. This is one reason why typeof() can come in handy.

Let's take a look at a quick example contrasting JavaScript with Java:

Java	JavaScript

`int age;`
`age =  38;`
`age = "thirty-eight";`
	`let age`
	`age = 38`
	`age = "thirty-eight"`

If we tried to run the Java code, we'd get an error stating that the types are incompatible. In Java, *variables* have a type. When we run the JavaScript code, however, everything's just fine. In JavaScript, *values* have a type.

It's also important to know that JavaScript is *weakly typed*, which means that implicit conversion between data types is allowed in most cases. If we recall the loose and strict equality operators from Chapter 3, *Nitty-Gritty Grammar*, weak typing is why current best practices specify using strict equality checking wherever possible.

If we take a look at a few languages on the scales of strong/weak and dynamic/static, we can plot the languages on an axis like this:

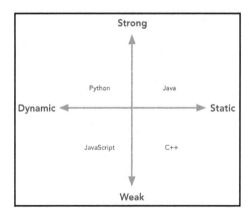

Figure 4.1 – Axes of typing

JavaScript style usually advocates for descriptive names rather than shorthand names. One of the reasons why this is acceptable is that often, JavaScript code is *minified* before it goes to production. It's not exactly like compiling, but it does condense whitespace and rename variables to be condensed. We'll discuss some of these build processes in `Chapter 16`, *Enter Webpack,* when we discuss webpack.

OK, so JavaScript is dynamically and weakly typed. What does that mean in practice? The short answer is this: be careful! It's very easy to get your types mixed up in a comparison operator or, even worse, accidentally cast a variable to a different type. It gives us much more flexibility as we're writing our program, but it can also be a curse. Some developers like to use Hungarian notation (`https://frontstuff.io/write-more-understandable-code-with-hungarian-notation`) to help differentiate variable types, but this isn't common practice in JavaScript. Probably the best way to help yourself and your colleagues keep your types correct is to be explicit with your variable names.

Exploring data types

Let's take a deeper dive into the primitive data types because they'll be crucial to our work in JavaScript. We not only need to know *what* we're using, but the *why* is also important. Our **primitives** are the building blocks of the rest of the language: Booleans, numbers, and strings. The rest of JavaScript is built upon these primitive data types. We'll start with Booleans.

Booleans

The **Boolean** is possibly the simplest and most universal data type since it's inherently tied to the 1s and 0s of binary logic. In JavaScript, a Boolean is written simply as `true` or `false`. It's not recommended to use `1` or `0` for Boolean values, as they'll be interpreted as numbers and thus fail strict equality. Boolean values are a specific data type, as opposed to in Python, where, at the core of the language, Boolean inherits from a number.

Remember in `Chapter 3`, *Nitty-Gritty Grammar*, where we learned that almost everything in JavaScript is an object? The same applies to Booleans. As you can see in the following screenshot, if you bring up the JavaScript console in your browser, chances are that it will autocomplete for you to see the methods available to you for Booleans:

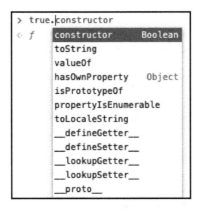

Figure 4.2 – Boolean autocomplete in Chrome

Now, I doubt any of these methods are particularly useful to you, but it's a handy way to check what methods are available to you for a given variable.

Booleans only get us so far—it's time to look at **numbers** next.

Numbers

JavaScript doesn't have a conception of different types of numbers such as integers, floats, or doubles—everything is simply a number. All the basic arithmetic methods are built-in, and the `Math` object provides the rest of the functionality you'd expect to find built into a programming language. Here's an example:

```
let myNumber = 2.14
myNumber = Math.floor(myNumber) // myNumber now equals 2
```

You can also use scientific notation, as shown:

```
myNumber = 123e5  // myNumber is 12300000
```

Numbers in JavaScript are not just any old numbers, but are, inherently, floats. To be technical, they are stored as double-precision floating-point numbers following the international IEEE 754 standard. However, this does lead to a couple of…interesting…quirks. Keep this in mind if you get strange results, such as in the following screenshot from the JavaScript console:

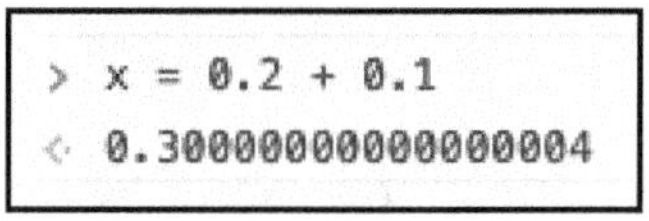

```
> x = 0.2 + 0.1
< 0.30000000000000004
```

Figure 4.3 – Floating-point precision error

A rule of thumb is to think about what precision you want your calculations to be. You can use the `toPrecision()` method of a number to specify your precision and then the `parseFloat()` function, like so:

```
let x = 0.2 + 0.1 // x = 0.30000000000000004
x = parseFloat(x.toPrecision(1)) // x = 0.3
```

`toPrecision()` returns a string, which may seem counter-intuitive at first, but there's a good reason for it. Let's say you needed your number to have two decimal places (for example, to display dollars and cents). If you used `toPrecision()` on a number and it returned a number, if you were to do more calculations on the integer, it would render only the integer unless you manipulated the decimal places as well. There is some method to the madness.

Next up: **strings**. We need to add some content to our programs!

Strings

Ah, the venerable string data type. It has some of the basics you'd expect, such as a `length` property and the `slice()` and `split()` methods, but two that always trip me up are `substr()` and `substring()`:

```
"hello world".substr(3,5) // returns "lo wo"
"hello world".substring(3,5) //  returns "lo"
```

The difference between the two methods is that the first one specifies `(start, length)`, while the second specifies `(start, end index)`. A handy way to remember the difference is that `.substring()` has an "i" in the name, correlating with index—the place in the string at which to stop.

A new addition in ES6 that makes our life easier is template literals. Take a look at this log:

```
const name = "Bob"
let age = 50
console.log("My name is " + name + " and I am " + age + " years old.")
```

It will work, but it's a little clunky. Let's use template literals:

```
console.log(`My name is ${name} and I am ${age} years old.`)
```

There are two important things to note in this example:

- The string starts and ends with a backtick, not a quote.
- The variables to insert are encased in `${ }`.

Template literals are handy, but they're not required. As you research code online when you run into problems, you will definitely see examples of the older, string-concatenated way of writing. However, keep in mind that this is an option for you.

Let's try our hand at an exercise!

Exercise – a basic calculator

With our knowledge of Booleans, numbers, and strings, let's build a basic calculator. Begin by cloning the repository at `https://github.com/PacktPublishing/Hands-on-JavaScript-for-Python-Developers/tree/master/chapter-4/calculator/starter-code`.

You can safely ignore the HTML and CSS for the most part, but a readthrough of the HTML will help. Let's take a look at the JavaScript:

```
window.onload = (() => {
  const buttons = document.getElementsByTagName('button')
  const output = document.getElementsByTagName('input')[0]
  let operation = null
  let expression = firstNumber = secondNumber = 0

  output.value = expression

  const clickHandler = ((event) => {
    let value = event.target.value

    /** Write your calculator logic here.
        Use conditionals and math to modify the output variable.

        Example of how to use the operators object:
```

```
            operators['='](1, 2) // returns 3

        Expected things to use:
          if/else
          switch() - https://developer.mozilla.org/en-
          US/docs/Web/JavaScript/Reference/Statements/switch
          parseFloat()
          String concatenation
          Assignment
      */

   })

   for (let i = 0; i < buttons.length; i++) {
     buttons[i].onclick = clickHandler
   }

   const operators = {
     '+': function(a, b) { return a + b },
     '-': function(a, b) { return a - b },
     '*': function(a, b) { return a * b },
     '/': function(a, b) { return a / b }
   };
 })
```

This is not an easy exercise for a beginner to JavaScript, so don't be afraid to check out the solution code and reverse-engineer it: https://github.com/PacktPublishing/Hands-on-JavaScript-for-Python-Developers/tree/master/chapter-4/calculator/solution-code.

Next, let's explore **arrays** and a new addition to ES6: **sets**.

Arrays and sets

Any programming language has some conception of an array or a *collection of items* that all share some common features or use. JavaScript has a few of them: **arrays** and **sets**. Both of these structures contain items, and in many ways, they are similar in usage, too, in that they can be enumerated, iterated over, and displayed for purposes of logical construction.

Let's first look at arrays.

Arrays

Arrays can contain different data types. This is a fully viable array:

```
const myArray = ['hello',1,'goodbye',null,true,1,'hello',{ 0 : 1 }]
```

It contains strings, numbers, Booleans, `null`, and an object. This is fine! While in practice you may not be mixing data types, there's nothing preventing you from doing so.

There is a quirk about using `typeof()` on arrays: since they're not true primitives, `typeof(myArray)` will return `object`. You should keep that in mind as you write JavaScript.

As we saw before in `Chapter 3`, *Nitty-Gritty Grammar*, `.push()` and `.pop()` are two of the most useful array methods, to add and remove items from an array, respectively. There are a good number of other methods available, though. Let's take a look at a few.

To create an array, we can do so as in the previous code or simply as `const myArray = []`. Now, while we can modify the values inside an array, we can declare it as a `const` because, for most purposes, we wouldn't want to let the program completely redefine it. We can still manipulate the values inside the array; we just don't want to destroy and recreate it. Let's continue with our array with items from the preceding example:

```
myArray.push('goodbye') // myArray is now
['hello',1,'goodbye',null,true,1,'hello',{ 0 : 1 }, 'goodbye']
myArray[3] // equals null
```

Remember that arrays are zero-indexed, so our counting begins at 0.

To remove an element from the end of an array, we use `.pop()`, like so:

```
let myValue = myArray.pop() // myValue = 'goodbye'
```

To remove an object from the beginning of an array, use `.shift()`, as follows:

```
myValue = myArray.shift() // myValue now equals 'hello'
```

Be aware that all of these methods introduced so far mutate the original array directly. `.pop()` and `.shift()` return the excised value, not the array itself. This distinction is important because not all array methods behave this way. Let's take a look at `slice` and `splice`:

```
myValue =  myArray.slice(0,1) // myValue equals 1, and myArray is unchanged
myValue = myArray.splice(0,1,'oh no') // myValue = 1, and myArray equals
['oh no', 'goodbye', null, true, 1, 'hello',{ 0 : 1 }]
```

You can look up the parameters for these two methods on the **MDN Web Docs** site. For the purposes of introducing these methods, just be aware that the behavior of methods on a variable can change from mutational to stable.

Sets are closely related to arrays but have some slight differences. Let's take a look.

Sets

Sets are a compound data type introduced in ES6. A set is an array with duplicates removed and that prohibits adding duplicates. Try the following code:

```
const myArray = ['oh no', 'goodbye', null, true, 1, 'hello',{ 0 : 1 }]
myArray.push('goodbye')
console.log(myArray)

const mySet = new Set(myArray)
console.log(mySet)

mySet.add('goodbye')
console.log(mySet)
```

`myArray` will have a length of 8, while `mySet` will have a length of 7—even *after* trying to add `'goodbye'`. JavaScript's `.add()` method of a set will first test to be sure a unique value is being added. Note the `new` keyword and the capitalization of the data type; this is not unique to creating sets, but it is important. In ES5 and before, it was common practice to declare new variables this way, but that practice is now considered legacy except for in a few instances.

 There's a common introductory-level JavaScript question in interviews that asks you to deduplicate an array. You can use a **set** to do this in one fell swoop instead of iterating through the array and checking each value.

While there are many possible solutions to deduplicating an array without using sets, let's take a look at a fairly basic example that uses the `.sort()` method. As you can expect from the name, this method will sort an array in ascending order. This approach is best used if you know the array will contain the same data type of strings or numbers.

Consider the following array:

```
const myArray = ['oh no', 'goodbye', 'hello', 'hello', 'goodbye']
```

We know a deduplicated, sorted array should look as follows:

```
['goodbye', 'hello', 'oh no']
```

We can test it as follows:

```
const mySet = new Set(myArray.sort())
```

Now, let's try it without using sets. Here is one approach that uses a deduplicating function:

```
const myArray = ['oh no', 'goodbye', 'hello', 'hello', 'goodbye']

function unique(a) {
  return a.sort().filter(function(item, pos, ary) {
    return !pos || item != ary[pos - 1]
  })
}

console.log(unique(myArray))
```

Go ahead and take a look: https://github.com/PacktPublishing/Hands-on-JavaScript-for-Python-Developers/blob/master/chapter-4/deduplicate/index.html.

What's the output? We should get an array with a length of 3, as follows:

```
["goodbye", "hello", "oh no"]
```

The raw approach is a bit more complex, right? Sets are a much more user-friendly way of deduplicating an array. Objects are another type of collection in JavaScript. As promised, here's a deeper dive into them.

Objects and JSON

Objects! Objects are at the core of JavaScript. As mentioned before in Chapter 3, *Nitty-Gritty Grammar*, almost everything in JavaScript is, at its core, an object. Objects may be intimidating at first, but they're easy enough to grasp in theory:

Here's the skeleton of an object:

```
const myObject = { key: value }
```

An object is a collection of *key/value pairs*. They're useful for many reasons, especially to contain and organize data. Let's look at the example of Captain Picard from the Chapter 3, *Nitty-Gritty Grammar*:

```
const captain = {
  "name": "Jean-Luc Picard",
  "age": 62,
  "serialNumber": "SP 937-215",
  "command": "NCC 1701-D",
  "seniorStaff": ['Riker','Data','Worf', 'Troi']
}
```

As we saw, we can use dot notation to access the properties of an object, like so:

```
captain.command // equals "NCC 1701-D"
```

We can also use other data types as values, as with `captain.seniorStaff`.

As with everything else, objects also have their own methods. One of the most handy ones is `.hasOwnProperty()`:

```
console.log(captain.hasOwnProperty('command')) // logs true
```

Now, let's try our array deduplication again, but this time, let's leverage objects to create a hashmap:

```
const myArray = ['oh no', 'goodbye', 'hello', 'hello', 'goodbye']

function unique_fast(a) {
  const seen = {};
  const out = [];
  let len = a.length;
  let j = 0;
  for (let i = 0; i < len; i++) {
    const item = a[i];
    if (seen[item] !== 1) {
      seen[item] = 1;
      out[j++] = item;
    }
  }
  return out;
}

console.log(unique_fast(myArray))
```

Let's take a peek: `https://github.com/PacktPublishing/Hands-on-JavaScript-for-Python-Developers/blob/master/chapter-4/deduplicate/hashmap.html`. Now, it's not immediately obvious, but this approach is nearly twice as fast as the deduplication method we explored before. Why? In short, an object's values can be accessed immediately in O(1) time versus iterating through the whole array in O(n). If you're unfamiliar with Big O notation, which is a fuzzy way to calculate code complexity, here's a good primer: `https://www.topcoder.com/blog/big-o-notation-primer/`.

Let's take a side-by-side comparison of the two approaches with an array of length 24,975.

The first implementation, `https://github.com/PacktPublishing/Hands-on-JavaScript-for-Python-Developers/blob/master/chapter-4/deduplicate/large.html`, will result in a time between 5 and 8 milliseconds (your mileage may vary).

However, by using a hashmap with an object, we can reduce our runtime by at least a few milliseconds: `https://github.com/PacktPublishing/Hands-on-JavaScript-for-Python-Developers/blob/master/chapter-4/deduplicate/large_hashmap.html`.

Now, a few milliseconds may seem insignificant (and impossible to distinguish with the eye) but think about an operation that needs to run over and over against datasets of a similar length. The savings add up.

You can take a peek at `https://stackoverflow.com/a/9229821/2581282` for some more ideas and explanations for this problem.

Next, we're going to examine something that makes JavaScript...well, *JavaScript!* Its notion of inheritance and classes is quite different from other languages. Let's dive in.

Prototypal inheritance

Inheritance in JavaScript really is one of its major strengths. Instead of classical class-based inheritance, JavaScript uses **prototypal** inheritance. (Protip: it's pronounced *pro-to-TYPE-al* not *pro-to-TYPICAL*.) That's because it uses the object's prototype as a template. Do you remember previously when we worked with the methods of a string and a number in the console and found a bunch of methods available to us even on a simple data type? Well, we can go further than that.

Fundamental to the concept of prototypal inheritance in JavaScript is the prototype chain, which tells us what we have access to in terms of methods. Let's take a look at a diagram:

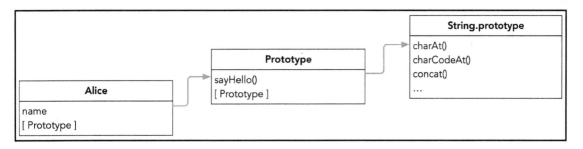

Figure 4.4 – The prototype chain

So, what does this mean? Consider `Alice`: we can see that this variable is a string as it descends from the `String` prototype. So, translated into code, we can say the following:

```
const Alice = new String()
Alice.name = "Alice"
console.log(Alice.name)
```

What will we get in the console? Simply `Alice`. We've given the *property* of `name` to our `Alice` string object. Now, let's take a look at this mysterious `sayHello()` method in the prototype. What do you think would happen if we executed the following?

```
Alice.sayHello()
```

If you guessed that we would get an undefined error on the `sayHello()` function, you'd be correct. We haven't defined it yet.

Here's how we do so by modifying the `String` prototype:

```
String.prototype.sayHello = function() {
  console.log(`My name is ${this.name}.`)
}
const Alice = new String()
Alice.name = "Alice"
Alice.sayHello()
```

Now, in our console, we will get `My name is Alice`. OK, so what happened?

By directly modifying the `String` prototype and adding a `sayHello()` method, we can use this method on any string and access its properties. Just like we used dot notation before, we can use the `this` keyword to refer to properties of the object within which we are working. Consequently, `this.name` works inside our prototype and equals `Alice.name`.

Now, you may be thinking *this seems a little dangerous*. We're modifying a base data type, and if we attempted to call `.sayHello()` on a string that *doesn't* have a `name` property, we'll get a nice huge error. You're correct! There's a better way to do this that still utilizes the concept of prototypal inheritance. Take a peek at this:

```
function Person(name) {
  this.name = name

  this.sayHello = function() {
    console.log(`My name is ${this.name}.`)
  }
}

const Alice = new Person('Alice')
const Bob = new Person('Bob')

Alice.sayHello()
Bob.sayHello()
```

As we would expect, we get `My name is Alice.` and `My name is Bob.`. We didn't need to define `sayHello()` twice; rather, `Alice` and `Bob` *inherited* the method from `Person`. Efficiency!

Now we're going to talk about Jason. Jason who? No, no, the object-based data structure called **JSON** is what we're going to examine next.

JSON

JSON (pronounced *jay-sohn* or *jason*) stands for **JavaScript Object Notation**. If you've seen it in the field before, you may know that it's often used as a convenient transfer format with APIs. We'll discuss APIs a little more in a bit, but for now, let's understand what JSON is and why it's useful.

Let's take a look at what it looks like. We'll be using the **Star Wars API (SWAPI)** (https:// swapi.dev) as a convenient read-only API. Take a peek at this example result: https:// swapi.dev/api/people/1/?format=json:

```
{"name":"Luke
Skywalker","height":"172","mass":"77","hair_color":"blond","skin_color":"fair","eye_color":"blue","bir
th_year":"19BBY","gender":"male","homeworld":"http://swapi.dev/api/planets/1/","films":
["http://swapi.dev/api/films/1/","http://swapi.dev/api/films/2/","http://swapi.dev/api/films/3/","http
://swapi.dev/api/films/6/"],"species":[],"vehicles":
["http://swapi.dev/api/vehicles/14/","http://swapi.dev/api/vehicles/30/"],"starships":
["http://swapi.dev/api/starships/12/","http://swapi.dev/api/starships/22/"],"created":"2014-12-
09T13:50:51.644000Z","edited":"2014-12-20T21:17:56.891000Z","url":"http://swapi.dev/api/people/1/"}
```

Figure 4.5 – SWAPI people instance

One of the great things about JSON is that it's fairly legible, as it doesn't have lots of nodes and formatting like XML. However, in its raw format, as in the preceding screenshot, it's still a jumble. Browsers have great tools to parse JSON into a legible tree. Take a minute to find and install one for your browser and then visit the previous API call. Now, your response should be formatted as in the following screenshot:

```
{
    name: "Luke Skywalker",
    height: "172",
    mass: "77",
    hair_color: "blond",
    skin_color: "fair",
    eye_color: "blue",
    birth_year: "19BBY",
    gender: "male",
    homeworld: "http://swapi.dev/api/planets/1/",
  - films: [
        "http://swapi.dev/api/films/1/",
        "http://swapi.dev/api/films/2/",
        "http://swapi.dev/api/films/3/",
        "http://swapi.dev/api/films/6/"
    ],
    species: [ ],
  - vehicles: [
        "http://swapi.dev/api/vehicles/14/",
        "http://swapi.dev/api/vehicles/30/"
    ],
  - starships: [
        "http://swapi.dev/api/starships/12/",
        "http://swapi.dev/api/starships/22/"
    ],
    created: "2014-12-09T13:50:51.644000Z",
    edited: "2014-12-20T21:17:56.891000Z",
    url: "http://swapi.dev/api/people/1/"
}
```

Figure 4.6 – SWAPI formatted

That's a lot more legible now. Say hi to Luke Skywalker!

One of the design decisions made by the authors of this API was to include in each result only the unique data of the resource in the result. For example, for homeworld, it doesn't spell out "Tatooine" but rather provides a **URI (Uniform Resource Identifier)** for a *planet* resource. We can see that homeworld and its data are key-value pairs, just like other objects, films is an array of strings, and the entire dataset is an object with curly braces at the beginning and end. That's all there is to JSON: properly formatted JavaScript objects.

Now it's time to dive into a couple of pieces of information about how the internet works to inform our use of JavaScript, APIs, and the greater web in general.

The HTTP verbs

Let's take a quick look at the HTTP verbs that allow us to communicate back and forth with APIs:

HTTP Verb	CRUD Equivalent
POST	Create
GET	Read
PUT	Update/Replace
PATCH	Update/Modify
DELETE	Delete

While the actual verbs used in an API depend on the API's design, these are the standard REST terms that many APIs today use. **REST** stands for **REpresentational State Transfer** and is a standard description of how to format APIs. Now, REST or RESTful APIs don't always have to communicate with JSON—REST is agnostic to format. Let's take a look at API calls in practice.

API calls from the frontend – Ajax

Ajax (also spelled AJAX) stands for **Asynchronous JavaScript and XML**. These days, however, you're more likely to work with JSON than XML, so the name is a bit misleading. On to code: take a look at `https://github.com/PacktPublishing/Hands-on-JavaScript-for-Python-Developers/blob/master/chapter-4/ajax/swapi.html`. Open this locally, and in your developer tools, you should see a JSON object as follows:

```
{"name":"Luke
Skywalker","height":"172","mass":"77","hair_color":"blond","skin_color":"fair","eye_color":"blue","bir
th_year":"19BBY","gender":"male","homeworld":"http://swapi.dev/api/planets/1/","films":
["http://swapi.dev/api/films/1/","http://swapi.dev/api/films/2/","http://swapi.dev/api/films/3/","http
://swapi.dev/api/films/6/"],"species":[],"vehicles":
["http://swapi.dev/api/vehicles/14/","http://swapi.dev/api/vehicles/30/"],"starships":
["http://swapi.dev/api/starships/12/","http://swapi.dev/api/starships/22/"],"created":"2014-12-
09T13:50:51.644000Z","edited":"2014-12-20T21:17:56.891000Z","url":"http://swapi.dev/api/people/1/"}
```

Figure 4.7 – SWAPI Ajax result

Congratulations! You've made your first Ajax call! Let's break down the following code:

```
fetch('https://swapi.co/api/people/1/')
  .then((response) => {
    return response.json()
  })
```

```
  .then((json) => {
    console.log(json)
  })
```

`fetch` is a fairly new API in ES6 that essentially replaces the older, more complicated way of making Ajax calls with `XMLHttpRequest`. As you can see, the syntax is fairly concise. What might not be obvious is the role that the `.then()` functions play—or even what they are.

`.then()` is an example of a Promise. We won't discuss Promises at length right now, but the basic premise hinges upon the asynchronous part of JavaScript. Essentially, a Promise says: "Execute this code, and I promise that at a later time, I'll provide you with more data. Don't block code execution here."

Open `https://github.com/PacktPublishing/Hands-on-JavaScript-for-Python-Developers/blob/master/chapter-4/ajax/swapi-2.html` locally in your browser. You should see **Loading Data...** for a quick second, and then the JSON displayed. You can use your browser's developer tools to throttle your internet connection to see this in action.

Here's the JavaScript code:

```
fetch('https://swapi.co/api/people/1/')
  .then((response) => {
    return response.json()
  })
  .then((json) => {
    document.querySelector('#main').innerHTML = JSON.stringify(json)
  })
document.querySelector('#headline').innerHTML = "Luke Skywalker"
```

Don't worry too much about the `document.querySelector` lines—we'll cover those in detail in `Chapter 6`, *The Document Object Model (DOM)*. For now, just understand that they're used to place information in an HTML document. Let's throttle our connection using the developer tools down to Slow 3G or similar. When we refresh, we should see a flash of **Awaiting Headline...**, then **Luke Skywalker**, followed by **Loading Data...**, and *then*, a few seconds later, the JSON as text.

So, how does this work? The line of code to change **Awaiting Headline...** to **Luke Skywalker** comes after the Ajax call. So why does the headline change *before* the data section? The answer is *Promises*.

Using `fetch`, we establish that we're inherently using asynchronous data, and thus the `.then()` statements tell us what we can do *after* the Promised statement resolves. It frees the program to continue on to other parts of the program. In fact, we could make multiple fetch calls that may return at various times, and still not block our user's use of the program. Asynchronicity is a fundamental concept when working with modern JavaScript, so take your time understanding it.

Next, let's get some experience with actually *using* APIs! It's time to really get our hands dirty and interact with not just local code but also external code.

SWAPI lab

Let's get some hands-on practice with this API. What we're going to do is fairly inelegant for the moment, but it will show us how to use asynchronous behavior to our advantage.

You should expect to see something like this:

Hello
Hello! My name is Luke Skywalker and I'm from Tatooine. I've been in The Force Awakens, Return of the Jedi, A New Hope, The Empire Strikes Back, Revenge of the Sith, and I'm a Jedi.

Figure 4.8 – SWAPI Promises result

Keep in mind that since we're using Promises and have to iterate over the `films` array, the order of films may vary. You can choose to order them by film number if you wish.

This lab will require nested Promises and some syntaxes we haven't covered yet, so give yourself plenty of time to experiment if you'd like to do this lab:

- Starter code: https://github.com/PacktPublishing/Hands-on-JavaScript-for-Python-Developers/tree/master/chapter-4/ajax-lab/starter-code
- Solution code: https://github.com/PacktPublishing/Hands-on-JavaScript-for-Python-Developers/tree/master/chapter-4/ajax-lab/solution-code

As with any lab, keep in mind that the solution code won't match your code, but is meant as a resource for thought processes.

Summary

Data is at the heart of every program, and your JavaScript programs are no different:

- JavaScript is loosely typed, which means variable types can mutate if needed.
- Booleans are simple true/false statements.
- Numbers are non-differentiated between integers, floats, or other types of numbers.
- Arrays and sets can contain a lot of data and make organizing our data easier.
- Objects are key-value pairs that efficiently store data for O(1) retrieval.
- API calls are actually not that scary!

We've taken a closer look at data types, APIs, and JSON. What we've discovered is that data is very flexible in JavaScript, up to and including manipulating the prototypes of the objects themselves. Taking a look at JSON and APIs, we've successfully used `fetch()` to perform our first API calls.

In the next chapter, we'll dive further into writing JavaScript to make a more interesting application, as well as understanding the details of how to construct one!

Questions

For the following questions, select the correct option:

1. JavaScript is inherently:
    1. Synchronous
    2. Asynchronous
    3. Both

2. A `fetch()` call returns a:
    1. `then`
    2. `next`
    3. `finally`
    4. Promise

3. With prototypal inheritance, we can (select all):
    1. Add methods to a base data type.
    2. Subtract methods from a base data type.
    3. Rename our data type.
    4. Cast our data into another format.

```
let x = !!1
console.log(x)
```

4. From the preceding code, what will be the expected output?
    1. `1`
    2. `false`
    3. `0`
    4. `true`

```
const Officer = function(name, rank, posting) {
  this.name = name
  this.rank = rank
  this.posting = posting
  this.sayHello = () => {
    console.log(this.name)
  }
}

const Riker = new Officer("Will Riker", "Commander", "U.S.S.
Enterprise")
```

5. In the preceding code, what's the best way to output `Will Riker`?
    1. `Riker.sayHello()`
    2. `console.log(Riker.name)`
    3. `console.log(Riker.this.name)`
    4. `Officer.Riker.name()`

Further reading

For more information on statically versus dynamically typed languages, you can refer to https://android.jlelse.eu/magic-lies-here-statically-typed-vs-dynamically-typed-languages-d151c7f95e2b.

To find out more about Hungarian notation, refer to https://frontstuff.io/write-more-understandable-code-with-hungarian-notation.

Section 2 - Using JavaScript on the Front-End

2

It's time to code! Let's put our theoretical knowledge of JavaScript into practice on the front-end by learning how to actually use it in practice on a page.

In this section, we will cover the following chapters:

Hello World! and Beyond: Your First Application

5

Ah, the venerable "Hello World!" script. While very simple, it's a good first test of any language. Let's do a little more than just saying hello, though; let's work with several small applications that we'll use to get our hands dirty. After all, there's more to programming than just theory. We'll take a look at a common problem presented in coding challenges, as well as understand *how* our programs are working.

The following topics will be covered in this chapter:

- I/O with the console and alert messages
- Working with input in a function
- Using objects as a datastore
- Understanding scope

Technical requirements

Clone or download the repository for this book from `https://github.com/PacktPublishing/Hands-on-JavaScript-for-Python-Developers` and prepare to look through the `Chapter-5` material.

I/O with the console and alert messages

So far, we've seen how JavaScript can output information to the user. Consider the following code:

```
const Officer = function(name, rank, posting) {
  this.name = name
  this.rank = rank
  this.posting = posting
  this.sayHello = () => {
    console.log(this.name)
  }
}

const Riker = new Officer("Will Riker", "Commander", "U.S.S. Enterprise")
```

Now, if we execute `Riker.sayHello()`, we will see the following in the console:

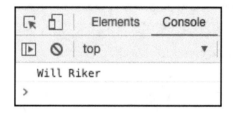

Figure 5.1 – Console output

Take a look for yourself in the `chapter-5` directory in the repository: `https://github.com/PacktPublishing/Hands-on-JavaScript-for-Python-Developers/blob/master/chapter-5/alerts-and-prompts/console.html`.

OK, great. We have some console output, but that's not a very efficient way to get output, as users don't typically have the console open. There is a convenient method for output that, while not practical for a fully fledged web application, is useful for testing and debugging purposes: `alert()`. Here's an example:

```
const Officer = function(name, rank, posting) {
  this.name = name
  this.rank = rank
  this.posting = posting
  this.sayHello = () => {
    alert(this.name)
  }
}
```

```
const Riker = new Officer("Will Riker", "Commander", "U.S.S. Enterprise")

Riker.sayHello()
```

Try running the preceding code from `https://github.com/PacktPublishing/Hands-on-JavaScript-for-Python-Developers/blob/master/chapter-5/alerts-and-prompts/alert.html` . What do you see?

Figure 5.2 – Alert message

Wonderful! We have one of those little annoying popups that you may have seen around the web. While they can be annoying when used improperly, they can be very handy when used appropriately.

Let's take a look at something similar that will give us *input* from the user (`https://github.com/PacktPublishing/Hands-on-JavaScript-for-Python-Developers/blob/master/chapter-5/alerts-and-prompts/prompt.html`):

```
const Officer = function(name, rank, posting) {
  this.name = name
  this.rank = rank
  this.posting = posting

  this.ask = () => {
    const values = ['name','rank','posting']

    let answer = prompt("What would you like to know about this officer?")
    answer = answer.toLowerCase()

    if (values.indexOf(answer) < 0) {
      alert('Value not found')
    } else {
      alert(this[answer])
    }
  }
}

const Riker = new Officer("Will Riker", "Commander", "U.S.S. Enterprise")

Riker.ask()
```

When you load the page, you will see a pop-up box with the input field. Enter `name`, `rank`, or `posting` to see the result. If you refresh and enter something other than those options, you should get a response of **Value not found**.

Ah! But let's also take a look at the following line:

```
answer = answer.toLowerCase()
```

Since this is frontend JavaScript, we don't know exactly what the user will enter, so we should plan for slight misformats. Data sanitization is a whole other topic, so for now, let's just agree that we can lowercase the whole string to match the expected values.

So far, so good. Now, let's take a look at how `answer` is used.

Working with input in a function

If we take a look at the preceding object, we'll see the following:

```
if (values.indexOf(answer) < 0) {
  alert('Value not found')
} else {
  alert(this[answer])
}
...
```

Since we're dealing with arbitrary input, the first thing we're doing is checking against our array of answers to see whether the property requested exists. If it does not, a simple error message is alerted. If it *is* found, then we can alert the value. If you remember from Chapter 3, *Nitty-Gritty Grammar*, object properties can be accessed via **dot notation** and **bracket notation**. In this case, we're working with a variable as the key, so we *can't* do this because it would be interpreted as the key. Thus, we use bracket notation to access the proper object value.

Exercise – Fibonacci sequence

For this exercise, construct a function to take a number. The end result should be the sum of numbers in the Fibonacci sequence (https://en.wikipedia.org/wiki/Fibonacci_number) up to the specified number you entered. The first few numbers of the sequence are [1, 1, 2, 3, 5, 8, 13, 21, 34, 55, 89]. Each number is the sum of the previous two numbers; for example, f[6] = 13 because f[5] = 8 and f[4] = 5, and thus f[6] = 8+5 = 13. You can use the starter code at https://github.com/PacktPublishing/Hands-on-JavaScript-for-Python-Developers/tree/master/chapter-5/fibonacci/starter-code. Don't worry too much about the most efficient algorithms for calculating the number; just be sure not to hardcode values and instead rely on the input variables and the formula.

Fibonacci sequence solution

Let's dissect one possible solution:

```
function fibonacci(num) {
  let a = 1, b = 0, temp

  while (num >= 0) {
    temp = a
    a = a + b
    b = temp
    num--
  }

  return b
}

let response = prompt("How many numbers?")
alert(`The Fibonacci number is ${fibonacci(response)}`)
```

Let's take a look at the lines outside of the function first. All we're doing is simply asking the user for the point in the sequence up to which they'd like to calculate. The response variable is then fed into the alert() statement as a parameter to fibonacci, which takes the argument of num. From that point forward, the while() loop executes on num, decrementing num as the value of b is incremented according to the algorithm, and then finally returns into our alert message.

So that's really all there is to it! Now, let's try a variant because we never know what our user will input. What happens if they enter a string instead of a number? We should accommodate this and at the very least present an error message.

Let's take a look at this solution:

```javascript
function fibonacci(num) {
  let a = 1, b = 0, temp

  while (num >= 0) {
    temp = a
    a = a + b
    b = temp
    num--
  }

  return b
}

let response = prompt("How many numbers?")

while (typeof(parseInt(response)) !== "number" ||
!Number.isInteger(parseFloat(response))) {
  response = prompt("Please enter an integer:")
}

alert(`The Fibonacci number is ${fibonacci(response)}`)
```

 You can find the solution on GitHub at `https://github.com/PacktPublishing/Hands-on-JavaScript-for-Python-Developers/tree/master/chapter-5/fibonacci/solution-code-number-check`.

If we dive into the `while()` loop, we'll see our type-matching magic. First of all, since `response` inherently comes in a string, we decide we don't want to trust type coercion, which is what our previous solution is doing. We use the `parseInt()` method to cast `response` directly into a number. Great! But that doesn't give us the security that our user entered an integer to begin with. Remember, JavaScript doesn't have a conception of `int` versus `float`, so we have to make some manipulations to ensure our input is an integer by using the negation of the `Number.isInteger` method. This ensures that our input is a valid integer.

As a prelude to heavier work with JSON, let's take a look at how we can use objects as a datastore.

Using objects as a datastore

Here's an interesting problem that I've seen in programming interviews, as well as the most efficient way to solve it. It has an expensive input time, but an O(1) *retrieval* time, which is generally considered a metric of success for algorithmic complexities when you can expect more reads than writes.

Exercise – multiplication

Consider the following code (`https://github.com/PacktPublishing/Hands-on-JavaScript-for-Python-Developers/tree/master/chapter-5/matrix/starter-code`):

```
const a = [1, 3, 5, 7, 9]
const b = [2, 5, 7, 9, 14]

// compute the products of each permutation for efficient retrieval

const products = { }

// ...

const getProducts = function(a,b) {
  // make an efficient means of retrieval
  // ...
}

// bonus: get an arbitrary key/value pair. If nonexistent, compute it and
store it.
```

So, what's the solution within the paradigm of using an object? Let's take a look, break it down, and then reverse-engineer our use of objects as a data store (*spoiler alert:* have you heard of NoSQL?).

Multiplication solution

Before we begin, let's break down the problem into two steps: given two arrays, we will first compute the products of each item in the arrays and store them in an object. Then, we will write a function to retrieve the product of two given numbers from the arrays. Let's take a look.

Step 1 – computing and storing

First of all, our `makeProducts` function will take the two arrays as its parameters. Using the `.forEach()` method of arrays, we'll iterate over each item in the first array, naming the values `multiplicant`:

```
const makeProducts = function(array1, array2) {
  array1.forEach( (multiplicant) => {
    if (!products[multiplicant]) {
      products[multiplicant] = { }
    }
    array2.forEach( (multiplier) => {
      if (!products[multiplier]) {
        products[multiplier] = { }
      }
      products[multiplicant][multiplier] = multiplicant * multiplier
      products[multiplier][multiplicant] = products[multiplicant]
        [multiplier]
    })
  })
}
```

Now, our end goal is to have an object that will tell us "the product of *x* and *y* is *z*." If we abstract this into using an object as a data store, we could arrive at a structure like this:

```
{
  x: {
    y: z
  },
  y: {
    x: z
  }
}
```

In this object structure, all we need to do to retrieve our calculation is specify x.y, which will be z. We also don't want to assume an order, so we do the reverse as well: y.z.

So, how do we construct this data object? Remember that we can use **bracket notation** with objects if we're not calling the literal key; here, we're using a variable:

```
if (!products[multiplicant]) {
  products[multiplicant] = { }
}
```

Our first step is to check whether the `multiplicant` key exists in our object (x, in our previous theoretical discussion). If it doesn't, set it to a new object.

Now, in our inner loop, let's do the same for the multiplier:

```
if (!products[multiplier]) {
    products[multiplier] = { }
}
```

Great! We have our keys set up for both x and y. Now, we just compute the product and store it in both locations, like so:

```
products[multiplicant][multiplier] = multiplicant * multiplier
products[multiplier][multiplicant] = products[multiplicant][multiplier]
```

Notice the decision to assign the reverse key value to the obverse key's value, as opposed to recalculating the product. Why would we do that? In fact, why are we going to all this trouble for a simple mathematical operation? Here's the reason: what if, instead of simple multiplication, we're doing a calculation *far* more complex? Maybe a calculation so complex it takes a full second or more to return? Now we can see that we want to reduce our time so that we only do the calculation once and then can read it repeatedly for optimal performance.

After constructing this function, we would execute it on our arrays:

```
makeProducts(a,b)
```

This is easy enough to invoke!

Step 2 – retrieval

Now, let's write our retrieval function:

```
const getProducts = function(a,b) {
  // make an efficient means of retrieval
  if (products[a]) {
    return products[a][b] || null
  }
  return null
}
```

If we look at this logic, first we ensure the first key exists. If it exists, we return either x.y or null if y does not exist. Objects are picky in that if you try to refer to a *value* of a *key* that doesn't exist, you'll get an error. Thus, we first need to existence-check our key. If the key exists *and* the key/value pair exists, return the computed value; otherwise, we return null. Notice the return products[a][b] || null short-circuit: this is an efficient way of saying "return the value or something else." If products[a][b] does not exist, it will respond with a falsy value, and the OR operation will take over. Efficient!

Take a look at the solution code for the answer to the bonus question. The same principles of existence-checking and calculation apply.

Understanding scope

Let's discuss scope for a while before we build a larger application. Simply put, scope defines when and where we can use a variable or a function. Scope in JavaScript is broken down into two discrete categories: local and global. If we look at our previous multiplication program, we can see that there are three variables outside any functions; they're hanging out at the root level of our program:

```
01: const a = [1, 3, 5, 7, 9]
02: const b = [2, 5, 7, 9, 14]
03:
04: // compute the products of each permutation for efficient retrieval
05:
06: const products = { }
07:
08: const makeProducts = function(array1, array2) {
09:     array1.forEach( (multiplicant) => {
10:         if (!products[multiplicant]) {
11:             products[multiplicant] = { }
12:         }
13:         array2.forEach( (multiplier) => {
14:             if (!products[multiplier]) {
15:                 products[multiplier] = { }
16:             }
17:             products[multiplicant][multiplier] = multiplicant *
                 multiplier
18:             products[multiplier][multiplicant] = products[multiplicant]
                 [multiplier]
19:         })
20:     })
21: }
22:
23: const getProducts = function(a,b) {
24:     // make an efficient means of retrieval
25:     if (products[a]) {
26:         return products[a][b] || null
27:     }
28:     return null
29: }
30:
31: makeProducts(a,b)
```

The variables in question are on lines 1, 2, and 6: a, b, and products, respectively. Great! That means that we can use them anywhere, such as on lines 10, 11, 14, 15, and more, as long as we use them after they're defined. Now, if we look closer, we also see we have some functions in the global scope: makeProducts and getProducts. Likewise, we can use them anywhere as long as they've already been defined.

OK, great—that makes sense since JavaScript is read top to bottom. But wait! If you remember from Chapter 3, *Nitty-Gritty Grammar*, a function declaration is hoisted to the top and thus can be used anywhere.

Let's refactor our program to utilize hoisting and abstract our math to be the theoretical long-running process. We'll also be using Promises as a great introduction to the concept. Before we go too far into it, it may be useful to read up on using Promises: https://developer.mozilla.org/en-US/docs/Web/JavaScript/Guide/Using_promises.

Take a look at index.js in https://github.com/PacktPublishing/Hands-on-JavaScript-for-Python-Developers/tree/master/chapter-5/matrix-refactored. We'll be breaking this down step by step.

First, open index.html in a browser. Make sure your console is open. After 2 seconds, you will see a simple message in the console: **9 x 2 = 18**. If you look at line 44 in index.js, you'll see that it's using getProducts to calculate the product of a[4] and b[0], which are 9 and 2, respectively. Great! So far, our functionality is the same with the addition of a perceived delay.

Let's start at the beginning:

```
1: const a = [1, 3, 5, 7, 9]
2: const b = [2, 5, 7, 9, 14]
3:
4: // compute the products of each permutation for efficient retrieval
5:
6: const products = {}
7:
```

OK, so far we have the same code. Now, what about our makeProducts function?

```
08: const makeProducts = async function(array1, array2) {
09:     const promises = []
10:     array1.forEach((multiplicant) => {
11:         if (!products[multiplicant]) {
12:             products[multiplicant] = {}
13:         }
14:         array2.forEach(async (multiplier) => {
15:             if (!products[multiplier]) {
```

```
16:                        products[multiplier] = {}
17:                    }
18:
19:                    promises.push(new Promise(resolve =>
                        resolve(calculation(multiplicant, multiplier))))
20:                    promises[promises.length - 1].then((val) => {
21:                        products[multiplicant][multiplier] = products[
                            multiplier][multiplicant] = val
22:                    })
23:                })
24:            })
25:        return promises
26: }
```

Hmm. OK, we have a few of the same pieces, but a few new pieces. First of all, let's consider `async`. This keyword, when used with a function, implies that the consumer of this function is to expect *asynchronous behavior* as opposed to JavaScript's generally top-down behavior. Before we dive into breaking down the new lines, 19–21, let's look at *why* this function is asynchronous by checking out our `calculation` function:

```
37: async function calculation(value1, value2) {
38:     await new Promise(resolve => setTimeout(resolve, 2000))
39:     return value1 * value2
40: }
```

Here's `async` again on line 37, and now we see a new keyword on line 38: `await`. `async` and `await` are one way to specify that we can work asynchronously: on line 38, we specify that we're waiting for this `promise` to **resolve** before continuing. What is our `promise` doing? Well, as it turns out, not a whole lot! It's simply using `setTimeout` to delay by 2,000 milliseconds. This delay is intended to simulate a long-running process, such as an Ajax call or a complex process that takes 2 seconds to complete (or even an indeterminate amount of time).

OK, great. So far, we've basically tricked the program into expecting a 2-second delay before continuing. Let's look at line 9: a new array called `promises`. Now, to get back to our idea of *scope*, you can notice that our array is defined *within* `makeProducts`. That means the variable only exists within the local scope of the function. As opposed to products, we can't access promises from outside this function. That's OK—we don't really need to. As a matter of fact, it's considered best practice to keep the number of variables defined in the global scope to a minimum.

Now, let's look at line 19, which looks a little more nuanced:

```
promises.push(new Promise(resolve => resolve(calculation(multiplicant,
multiplier))))
```

If we dissect this, we first see something familiar: we're pushing something onto our promises array. What we're pushing is a new Promise, similar to line 38, but in this case, we're not waiting for it in-line, but rather just saying "resolve this promise with the value of calculation() — whenever it happens." So far, so good. How about the next part?

```
20: promises[promises.length - 1].then((val) => {
21:     products[multiplicant][multiplier] = products[multiplier]
          [multiplicant] = val
22: })
```

Now, here's where some syntactic sugar comes into play: now that we have our promise in our array of promises, we access it with [promises.length - 1] because length returns the full length, starting from 1. The .then() clause is our magic: it says that once the promise is done, do something with the result. In this case, our *something* is to assign val to both variants of the product. Finally, on line 25, we return the array of promises.

Our getProducts function hasn't changed at all! The complexity of our retrieval function remains the same: efficient.

How about this?

```
42: makeProducts(a,b).then((arrOfPromises) => {
43:     Promise.all(arrOfPromises).then(() => {
44:         console.log(`${a[4]} x ${b[0]} = ${getProducts(a[4], b[0])}`)
              // 18
45:     })
46: })
```

We've seen .then before, so it's getting as its parameter the return value of makeProducts, which is the array of promises. Then, we can use .all() before .then to effectively say "when all the promises in arrOfPromises have resolved, then do the next function." That next function is to log our answer. You can add additional product checks after line 44; they will all return at the same time as line 44, as the delay in our "calculation" has already occurred.

Scope chains and scope trees

Diving further into scope, we have the idea of **scope chains** and **scope trees**. Let's consider the following example:

```
function someFunc() {
  let outerVar = 1;
  function zip() {
    let innerVar = 2;
  }
}
```

What variable(s) does `someFunc` have access to? What does `zip` have access to? If you guessed that `someFunc` has access to `outerVar` but `zip` has access to both `innerVar` and `outerVar`, you're correct. That's because both variables exist in the scope chain of `zip`, but only `outerVar` exists in the scope of `someFunc`. Clear as mud? Great. Let's look at some diagrams.

Take a look at the following code:

```
function someFunc() {
  function zip() {
    function foo() {
    }
  }
  function quux() {
  }
}
```

We can diagram a **scope tree** of our function from a top-down construction:

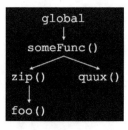

Figure 5.3 – Scope tree

What does this tell us? quux kind of lives off in its own little world inside someFunc. It would have access to someFunc's variables, but *not* to zip or foo. We can also look at it in the reverse with a **scope chain** to understand it from the bottom up:

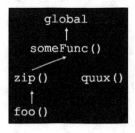

Figure 5.4 – Scope chain

In this example, we're taking a look at what foo has access to. From the bottom up, we can see its relationship to the other parts of the code.

Closures

Now, we'll get into **closures**, which is apparently a scary topic in JavaScript. However, the basic concept is approachable: a closure is simply a function inside another function with access to the scope chain of its parent function. In this case, it has three scope chains: its own, with variables defined within itself; global, with all the variables in the global scope accessible to it; and the parent function's scope.

Here's an example that we'll dissect:

```
function someFunc() {
    let bar = 1;

    function zip() {
        alert(bar); // 1
        let beep = 2;

        function foo() {
            alert(bar); // 1
            alert(beep); // 2
        }
    }
}
```

Which variables are accessible to which functions? Here's a diagram:

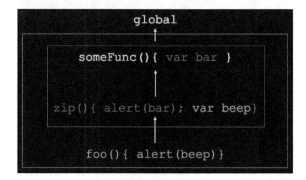

Figure 5.5 – Closures

Starting from the bottom up, `foo` has access to `beep` and `bar`, whereas `zip` has access only to `bar`. So far, so good, right? Closures are just a way to describe the scope that each nested function has available to it. There's nothing inherently scary about them.

A basic example of a closure in practice

Take a look at the following function:

```
function sayHello(name) {
  const sayAlert = function() {
    alert(greeting)
  }

  let greeting = `Hello ${name}`
  return sayAlert
}

sayHello('Alice')()
alert(greeting)
```

First, let's look at this interesting construction: `sayHello('Alice')()`. Since our `sayAlert()` function is the return value of `sayHello`, we first invoke `sayHello` with one pair of parentheses with our argument and then invoke its return value (the `sayAlert` function) with the second pair of parentheses. Notice how `greeting` is within the scope of `sayHello`, and when we invoke our function, we'll have an alert of **Hello Alice**. However, if we try to alert `greeting` by itself, we'll get an error. Only `sayAlert` has access to `greeting`. Likewise, if we tried to access `name` from outside the function, we'd get an error.

Summary

In order for our programs to be useful, they usually depend on input from the user or other functions. By scaffolding our programs to be flexible, we also need to keep in mind the idea of scope: when and where we can use a function or variable. We also took a look at how objects can be used to store data efficiently for retrieval.

Let's not forget closures, the seemingly complicated concept that is, in practice, just a way of describing scope.

In the next chapter, we'll explore the frontend more as we get into using the **Document Object Model** (**DOM**) and manipulating information on the page instead of only interacting with alerts and the console.

Questions

Consider the following code:

```
function someFunc() {
   let bar = 1;

   function zip() {
      alert(bar); // 1
      let beep = 2;

      function foo() {
         alert(bar); // 1
         alert(beep); // 2
      }
   }

   return zip
}


function sayHello(name) {
   const sayAlert = function() {
      alert(greeting)
   }

   const sayZip = function() {
      someFunc.zip()
   }
```

```
    let greeting = `Hello ${name}`
    return sayAlert
}
```

1. How would you get an alert of **Hello Bob**?
    1. `sayHello()('Bob')`
    2. `sayHello('Bob')()`
    3. `sayHello('Bob')`
    4. `someFunc()(sayHello('Bob'))`

2. What will `alert(greeting)` do in the preceding code?
    1. Alert **greeting**.
    2. Alert **Hello Alice**.
    3. Throw an error.
    4. None of the above.

3. How would we get an alert message of **1**?
    1. `someFunc()()`
    2. `sayHello().sayZip()`
    3. `alert(someFunc.bar)`
    4. `sayZip()`

4. How would we get an alert message of **2**?
    1. `someFunc().foo().`
    2. `someFunc()().beep.`
    3. We can't because it's not in the scope.
    4. We can't because it's not defined.

5. How can we change `someFunc` to alert **1 1 2** ?
    1. We can't.
    2. Add `return foo` after `return zip`.
    3. Change `return zip` to `return foo`.
    4. Add `return foo` after the `foo` declaration.

6. Given a correct solution to the preceding question, how would we actually get three alerts of **1, 1, 2** ?
    1. `someFunc()()()`
    2. `someFunc()().foo()`
    3. `someFunc.foo()`
    4. `alert(someFunc)`

Further reading

- MDN – closures: `https://developer.mozilla.org/en-US/docs/Web/JavaScript/Closures`
- *Understand JavaScript Closures with Ease*: `http://javascriptissexy.com/understand-javascript-closures-with-ease/`

6
The Document Object Model (DOM)

The **Document Object Model** (**DOM**) is the API exposed by the browser to allow JavaScript to communicate with HTML and, indirectly, CSS. Since one of JavaScript's main abilities is dynamically changing content on a page, we should know how to do that. Enter the DOM.

In this chapter, we will learn how to use this powerful API to read and change content on a page. I'm sure you've seen websites that change content without reloading the page. These programs use *DOM manipulation*, and we'll learn how to use it.

The following topics will be covered in this chapter:

- Selectors
- Properties
- Manipulations

Technical requirements

Be sure to have the `https://github.com/PacktPublishing/Hands-on-JavaScript-for-Python-Developers` repository handy and ready for use in the `Chapter-6` directory.

Using selectors

So far, we've only been using `console.log` and alerts and prompts to input and output information. While these methods are useful for testing, they're not exactly what you would use in everyday life. Most of the web applications that we use, from searching to email, use the DOM to interact with the user to get input and show information. Let's take a look at a small example: `https://github.com/PacktPublishing/Hands-on-JavaScript-for-Python-Developers/tree/master/chapter-6/hello`.

If you open the HTML in the browser, we see a very simple page:

Figure 6.1 Our basic page

If we click the button, we don't get an alert or a console message, but instead, we have this:

Figure 6.2 An in-page response to our click!

Yay! It's our first instance of **DOM manipulation**.

DOM manipulation explained

Let's look at the JavaScript that powered that amazing example:

```
document.querySelector('button').addEventListener('click', (e) => {
  document.querySelector('p').innerHTML = `Hello! It is currently ${
  new Date()}.`
})
```

The first thing to notice is that we're operating on the `document` object. `document` is JavaScript's conception of what the page in the browser consists of. Remember when I mentioned that the DOM is an API exposed by the browser? This is the vector by which you access the DOM: `document`.

Before we dissect the JavaScript, let's see how the DOM and HTML differ. Here's our HTML for our page:

```
<!DOCTYPE html>
<html lang="en" dir="ltr">

<head>
  <meta charset="utf-8">
  <title>Example</title>
</head>

<body>
  <p></p>
  <button>Click me!</button>
  <script src="index.js"></script>
</body>

</html>
```

If we use our console now to inspect **Elements** instead of **Console**, we'll see this:

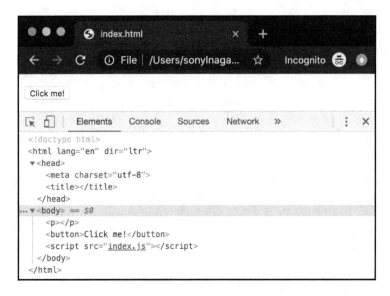

Figure 6.3 The DOM of our page

If you look closely and compare this screenshot with the preceding HTML, you won't really find any differences. However, now click the button and see what happens to the <p> tag:

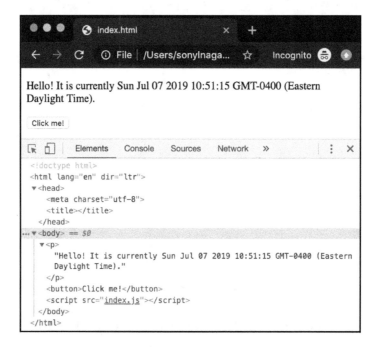

Figure 6.4 After clicking the button

Ah! Now we see a difference between the HTML and the DOM: the addition of text inside the paragraph tag. If we reload the page, poof, there goes our text and we're back to the beginning. So, what we're seeing is that nothing is changing *on disk*, only *in memory*. The DOM only lives in memory. You can experiment in the **Elements** view by changing values and even deleting whole **nodes**. A node is the DOM's reflection of an HTML tag. You may hear *node* and *tag* used interchangeably, but when working with JavaScript, it's a good habit to use *node* to be consistent with JavaScript's nomenclature, as we'll see in a bit.

Back to our JavaScript. So far, we've talked about document, which is the DOM's in-memory interpretation of the HTML. The method of document that we're using is a powerful one: .querySelector(). This method returns the *first* match to the argument we pass into the method. In this case, we're asking for button. Since there's only one button on the page, we can simply use the tag name. However, querySelector is more powerful than that, in that we can select based on CSS selectors as well. For example, say our button had a class on it like this:

```
<button class="clickme">Click me!</button>
```

We could then access the button like this:

```
document.querySelector('.clickme')
```

Notice the "." in front of `clickme`, just like a CSS selector. Similarly, when accessing an element with an ID, you would use "#".

Now that we have access to our button, we want to do *something* with it. In this case, *something* is to take action when the button is clicked. We do this by adding an **event listener**. We'll take a deeper dive into event listeners in `Chapter 7`, *Events, Event-Driven Design, and APIs*, so for now, let's just scratch the surface.

This is the structure of an event listener:

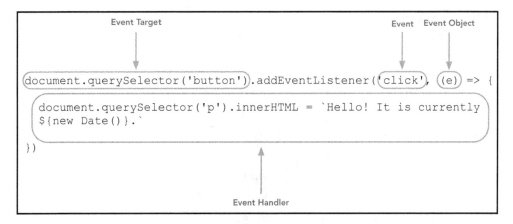

Figure 6.5 Event listener structure

First, our **event target** is the node upon which we want to listen; in this case, our target is the button. We then use the `.addEventListener()` method and assign the **event** of `click` to be the event for which we are listening. The second argument to our event listener is a function called the **event handler**. We can pass the actual **event object** to our handler. Event handlers don't have to be anonymous, as this one is, but it's common practice unless you need repeated functionality for several event types. Our handler is using `querySelector` again to target the p node and setting its `innerHTML` property to the string with our date.

A word about node properties: a node's *properties* are the DOM's in-memory representation of an HTML element's attributes. That means there are plenty of them: `className`, `id`, and `innerHTML`, just to name a few; we'll get into them more in a minute when we get to the *Properties* section. So all together, these lines of code tell the browser, "Hey, when this button is clicked, change the content of the p tag to be this string."

Now that we've taken a bird's eye view of this, let's dive into each of the pieces involved in making DOM manipulation work.

Using selectors

Let's consider a more complex page. We'll be opening a sample page and working with some of the elements provided to you:

1. Open `index.html` in `https://github.com/PacktPublishing/Hands-on-JavaScript-for-Python-Developers/tree/master/chapter-6/animals` in a browser:

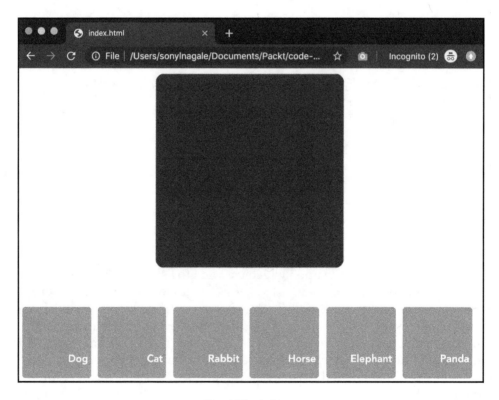

Figure 6.6 The animals page

2. If you hover over an orange button, it will turn turquoise and, when you click it, the black box at the top of the page will show the animal:

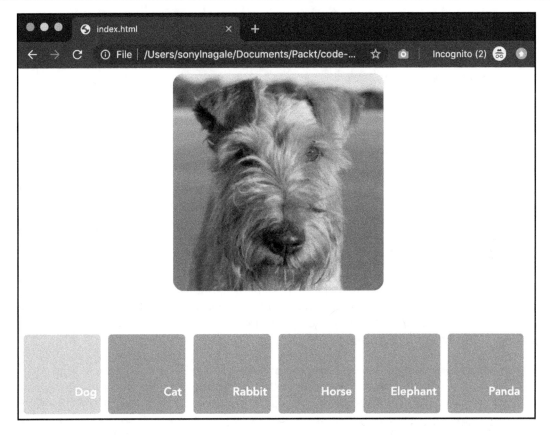

Figure 6.7 A selected animal

3. Take a minute to play around with the page and examine its behavior. Also, try hovering over the photo; what happens?

Now let's take a look at the JavaScript. Again, it's fairly simple, but we have a few new characters in our story:

```
01: const images = {
02:    'path': 'images/',
03:    'dog': 'dog.jpg',
04:    'cat': 'cat.jpg',
05:    'elephant': 'elephant.jpg',
06:    'horse': 'horse.jpg',
07:    'panda': 'panda.jpg',
08:    'rabbit': 'rabbit.jpg'
09: }
10:
```

```
11: const buttons = document.querySelectorAll('.flex-item');
12:
13: buttons.forEach((button) => {
14:   button.addEventListener('click', (e) => {
15:     document.querySelector('img').src =
          `${images.path}${images[e.target.id]}`
16:   })
17: })
18:
19: document.querySelector('#image').addEventListener('mouseover', (e) => {
20:   alert(`My favorite picture is ${e.target.src}`)
21: })
```

Lines 1-9 contain an object as a datastore. Great! We covered this usage in Chapter 5, *Hello World! and Beyond: Your First Application.*

Line 11 introduces a new way of using a selector: .querySelectorAll(). As mentioned previously, when we use .querySelector(), we'll get the *first* item that matches our query. This method will return an array of all matching nodes. Then, we can iterate over them on line 13 to give each of them a click handler. On line 15, we define *what happens* in our event handler: set the source of the only img node to be a concatenation of the path and image source from our data object.

But wait! What is e.target? We'll take a deeper dive into events in Chapter 7, *Events, Event-Driven Design, and APIs,* but for now, it's only important to know that e.target is *the DOM node of the event target.* So, in this example, we're iterating through all DOM nodes of the .flex-item class. On each node, we are assigning an event handler, thus e.target equals the DOM node and e.target.id equals its HTML attribute of id.

Fantastic. Let's take a look at line 19, where we're doing something similar, but this time using the CSS selector of id—image. Take a look at the HTML:

```
<div class="flex-header"><img id="image"/></div>
```

We see there is an ID of image on our tag, which means our DOM node will also have this ID. Now, when we move (or hover) over the image, we'll get an alert message stating the local path of the image file.

If you're not that fluent with CSS, right now you may be asking yourself: But where's the JavaScript to turn the orange boxes turquoise? Ha! Trick question! Let's look at lines 45-48 in the `style.css` file:

```
.flex-item:hover {
  cursor: pointer;
  background-color: turquoise;
}
```

If you notice the *pseudoclass* of `:hover` on the item, we see the CSS rules that change the cursor from an arrow to the hand (indicating clickability in most user interfaces) as well as the background color change. Surprise!

This is not a book on CSS; on the contrary, we're going to try to steer clear of too many style dependencies. However, it's important to note that often CSS allows us to make changes to some presentational aspects of HTML elements. But why do we care? After all, we're writing *JavaScript*. The answer is simple: computational expense. It's more *expensive* (that is, it takes more processing power) to modify an element via JavaScript than by CSS. If you're manipulating CSS attributes that don't require logic, use CSS where possible. However, if you need logic (such as stitching in variables to display an image, as in our example) then JavaScript is the correct choice.

Using other selectors

It's important to note that before `querySelector` and `querySelectorAll` were standardized as part of ES6 and HTML5, there were other selectors that were more prevalent, and you're certain to encounter them in the wild. Some of them include `getElementById`, `getElementsByClassName`, and `getElementsByTagName`. It's now considered a standard practice to use a variant of `querySelector`, but as with everything JavaScript, there is a caveat: technically, the `querySelector` methods are a tiny bit more expensive than the `getElement`-style methods. Usually, this expense is negligible when weighed against the power and flexibility of the `querySelector` methods, but it's something to keep in the back of your mind when dealing with large pages.

Now, let's take a look at *what* we can change after we've selected our elements. These are the **properties** of an element.

Properties

We've dealt with a few properties already: innerHTML of a node, src of an image, and id of a node. There is a vast array of properties available to us, so let's take a peek at how CSS marries with JavaScript.

Just for the sake of argument, let's change our Animals program to use JavaScript to change the background color of the target instead of CSS (https://github.com/PacktPublishing/Hands-on-JavaScript-for-Python-Developers/tree/master/chapter-6/animals-2):

```
const images = {
  'path': 'images/',
  'dog': 'dog.jpg',
  'cat': 'cat.jpg',
  'elephant': 'elephant.jpg',
  'horse': 'horse.jpg',
  'panda': 'panda.jpg',
  'rabbit': 'rabbit.jpg'
}

const buttons = document.querySelectorAll('.flex-item');

buttons.forEach((button) => {
  button.addEventListener('mouseover', (e) => {
    e.target.style.backgroundColor = 'turquoise'
  })
  button.addEventListener('click', (e) => {
    document.querySelector('img').src =
      `${images.path}${images[e.target.id]}`
  })
})

document.querySelector('#image').addEventListener('mouseover', (e) => {
  alert(`My favorite picture is ${e.target.src}`)
})
```

If we examine our mouseover handler, we can notice two things:

- The name of the event is mouseover, not hover. More on that later.
- We're modifying the style property of our target, but the name is backgroundColor, **not** background-color as it is in CSS.

The camel case rule of properties in CSS is also a standard in JavaScript. It may seem a bit counterintuitive to have a different set of nouns for JavaScript, but in practice, it's better because you don't have to use bracket notation and quotes to deal with the hyphens in property names (which would be interpreted as an invalid subtraction statement).

However, let's now run our program and hover over all of the boxes. Do you see a color change from one color to another, like this?

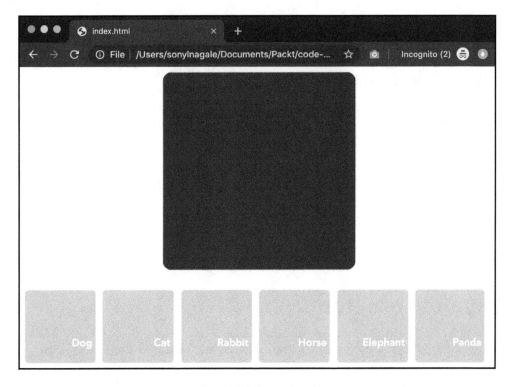

Figure 6.8 All the boxes are changed!

Yep, if you guessed that we didn't include a "reset" handler, you're right. We can do that with the `mouseout` event. However, do you see how it makes sense to use CSS when you can?

It's certainly not necessary to memorize all the various properties available to you on DOM nodes, but `id`, `className`, `style`, and `dataset` are probably the most useful.

What's this `dataset` property, you ask? You may not be familiar with data attributes in HTML, but they come in immensely handy. Consider this example from MDN:

```
<article id="electric-cars" data-columns="3" data-index-number="12314"
data-parent="cars"> ... </article>
```

The "data-" attributes are quite handy when your backend can insert markup into your HTML but is divorced from your JavaScript (as is almost always the case and, arguably, as your structure should be architected). To access `data-index-number` of `article`, we use this:

```
article.dataset.indexNumber // "12314"
```

Notice again our camel case and the new usage of `.dataset.`, not `data-`.

We now know enough to do some more exciting work with our elements. We can target elements with selectors and read the elements' attributes. Next, let's look at **manipulations**.

Manipulations

When working with the DOM via JavaScript, we can not only read but *manipulate* these properties. Let's get some practice in manipulating properties by making a small program: a sticky note creator.

Sticky note creator

We're going to make a sticky note creator that takes a color and a message and adds that colored box to the DOM with an ordinal number. Here's what our final product might look like:

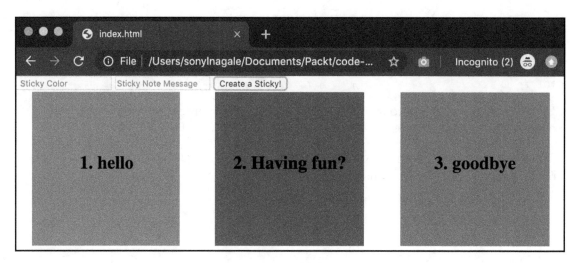

Figure 6.9 Final product

Take a look at the starter code: `https://github.com/PacktPublishing/Hands-on-JavaScript-for-Python-Developers/tree/master/chapter-6/stickies/starter-code`.

Your goal is to recreate this functionality. Here are two methods we haven't yet covered for you to research:

- `document.createElement()`
- `container.appendChild()`

Solution code

How did you do? Let's take a look at the solution code:

```javascript
const container = document.querySelector('.container') // set .container to
a variable so we don't need to find it every time we click
let noteCount = 1 // inital value

// access our button and assign a click handler
document.querySelector('.box-creator-button').addEventListener('click', ()
=> {
  // create our DOM element
  const stickyNote = document.createElement('div')

  // set our class name
  stickyNote.className = 'box'

  // get our other DOM elements
  const stickyMessage = document.querySelector('.box-color-note')
  const stickyColor = document.querySelector('.box-color-input')

  // get our variables
  const message = stickyMessage.value
  const color = stickyColor.value

  // blank out the input fields
  stickyMessage.value = stickyColor.value = ''

  // define the attributes
  stickyNote.innerHTML = `${noteCount++}. ${message}`
  stickyNote.style.backgroundColor = color

  // add the sticky
  container.appendChild(stickyNote)
})
```

OK! Some of these lines shouldn't be a mystery, but the most interesting ones are lines 7 (`const stickyNote = document.createElement('div')`) and 28 (`container.appendChild(stickyNote)`). As mentioned previously, these are the two methods you would need to research in order to finish this program. Line 7 is creating a DOM node—in memory! We can make our manipulations to it, such as adding content and style, and then on line 28 we're adding it to the DOM.

Summary

Yay, we've finally gotten into the DOM and manipulated it! Congrats on where you're at so far!

With JavaScript, we can now dynamically change what is on the page as opposed to only using alerts and console messages. Here's an overview of what we learned:

- `querySelector` and `querySelectorAll` are our gateways into the magical realm of the DOM.
- The DOM exists only in memory as a dynamic representation of where the HTML was when the page was loaded.
- Selectors for these methods will use CSS selectors; legacy methods will not.
- Properties of nodes can be changed, but the nomenclature varies.

In the next chapter, we'll work more with *events*. Events are at the heart of a JavaScript program, so let's learn about their structure and use.

Questions

Consider the following code:

```
<button>Click me!</button>
```

Answer the following question:

1. What is the correct syntax to select the button?
    1. `document.querySelector('Click me!')`
    2. `document.querySelector('.button')`
    3. `document.querySelector('#button')`
    4. `document.querySelector('button')`

Take a look at this code:

```
<button>Click me!</button>
<button>Click me two!</button>
<button>Click me three!</button>
<button>Click me four!</button>
```

Answer the following questions:

1. True or false: `document.querySelector('button')` will serve our needs to place a click handler on each button.
   1. True
   2. False

2. To change the text of the button from "Click me!" to "Click me first!", what should we use?
   1. `document.querySelectorAll('button')[0].innerHTML = "Click me first!"`
   2. `document.querySelector('button')[0].innerHTML = "Click me first!"`
   3. `document.querySelector('button').innerHTML = "Click me first!"`
   4. `document.querySelectorAll('#button')[0].innerHTML = "Click me first!"`

3. What method could we use to add another button?
   1. `document.appendChild('button')`
   2. `document.appendChild('<button>')`
   3. `document.appendChild(document.createElement('button'))`
   4. `document.appendChild(document.querySelector('button'))`

4. How can we change the class of the third button to `third`?
   1. `document.querySelector('button')[3].className = 'third'`
   2. `document.querySelectorAll('button')[2].className = 'third'`
   3. `document.querySelector('button[2]').className = 'third'`
   4. `document.querySelectorAll('button')[3].className = 'third'`

Further reading

For more information, you can refer to the following links:

- MDN: *Document Object Model (DOM)*: `https://developer.mozilla.org/en-US/docs/Web/API/Document_Object_Model`
- MDN: *Document.createElement()*: `https://developer.mozilla.org/en-US/docs/Web/API/Document/createElement`
- MDN: *Node.appendChild()*: `https://developer.mozilla.org/en-US/docs/Web/API/Node/appendChild`

7
Events, Event-Driven Design, and APIs

At the heart of a frontend application are *events*. JavaScript allows us to listen for and react to user and browser events to change content for the user in an intuitive fashion to create elegant user interfaces and experiences. We need to know how to use these packets of data that are thrown around. Browser events are our bread and butter—they allow us to have more than a static application and, instead, be dynamic! By understanding events, you'll be on your way to becoming a full JavaScript developer.

The following topics will be covered in this chapter:

- The event life cycle
- Capturing an event and reading its properties
- Using Ajax and events to populate API data
- Handling asynchronicity

Technical requirements

Be prepared to work with the code provided in the `Chapter-7` directory of the repository: `https://github.com/PacktPublishing/Hands-on-JavaScript-for-Python-Developers/tree/master/chapter-7`.

The event life cycle

When an event occurs in JavaScript, it doesn't simply happen and vanish—it goes through a *life cycle*. There are three phases to this life cycle:

- The **capture** phase
- The **targeting** phase
- The **bubbling** phase

Consider the following HTML:

```
<!doctype html>
<html>

<head>
  <title>My great page</title>
</head>

<body>
  <button>Click here</button>
</body>

</html>
```

We can visualize it as follows:

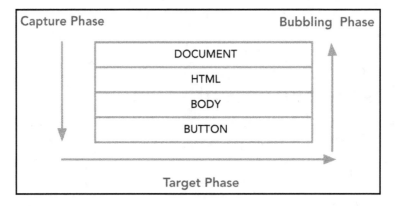

Figure 7.1 – The event life cycle

Now, there's something else that is important to consider when it comes to events: they don't just take effect on the exact target, but rather on the whole stack of objects. Before we describe what capturing, targeting, and bubbling entail, take a look at the following representation of our code:

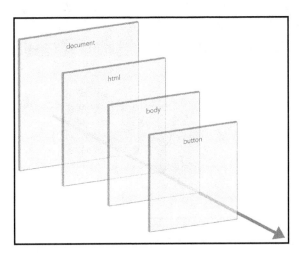

Figure 7.2 – Event layering

If we think about our page as a layer cake, we can see that this event (represented by the arrow) must pass through all the layers of our DOM to reach the button. This is our **capture** phase. When the button is clicked, an event is *dispatched* into the event flow. First, the event looks at the document object. It then travels through the layers of the DOM until it reaches the intended destination: the button.

Now that the event has reached the button, we begin the **targeting** phase. Whatever information the event is supposed to capture from the button will be gathered, such as the event type (such as a click or mouseover) and other details, such as the cursor's *X/Y* coordinates.

Finally, the event travels back through the layers to the document in the bubbling phase. The **bubbling** phase allows us to handle an event on *any* element by its parent elements.

Let's look at this in practice and play with our events a little bit. Find the following directory and open `index.html` in the browser—`https://github.com/PacktPublishing/ Hands-on-JavaScript-for-Python-Developers/tree/master/chapter-7/events`:

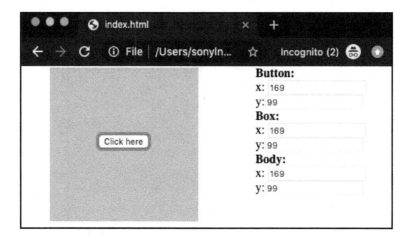

Figure 7.3 – Events playground

If we take a look at this page and play around with it for a few minutes, we see a few things:

- The *X/Y* coordinates on the right will change as we move our mouse on the page.
- When we open the console, it will display messages about our click event and at which *phase* it occurred.

Let's take a look at the code in `index.js` at `https://github.com/PacktPublishing/Hands-on-JavaScript-for-Python-Developers/blob/master/chapter-7/events/index.js`.

From lines 1 to 5, we're simply setting up a data object to map a numerical code to a string. So far, so good. Now, let's take a look at line 32, where it says `document.querySelector('html').addEventListener('click', logClick, true)`. This optional Boolean parameter is new to us right now; when put it into an event listener, it simply says "Let me listen in the *capture* phase." Thus, when we click anywhere on our page, we'll get a click event with the information **Click event triggered during capture phase at HTML.** And this event was previously handled at undefined because it's the very first encounter with this event. It hasn't yet bubbled or been targeted.

Let's keep dissecting this example in the next section to learn about these mysterious parts of the code.

Capturing an event and reading its properties

We'll continue working with our `events` playground code: https://github.com/PacktPublishing/Hands-on-JavaScript-for-Python-Developers/blob/master/chapter-7/events/index.js.

On lines 32–34, we registered three click event listeners, as shown:

```
document.querySelector('html').addEventListener('click', logClick, true)
document.querySelector('body').addEventListener('click', logClick)
document.querySelector('button').addEventListener('click', logClick)
```

As we discussed, the first one is listening in the capture phase because we've included the final Boolean parameter.

We also have three `mousemove` events on lines 16–29. Let's take a look at one of them:

```
document.querySelector('button').addEventListener('mousemove', (e) => {
  document.querySelector('#x').value = e.x
  document.querySelector('#y').value = e.y
})
```

I hope most of this makes sense—we're using a new event type of mousemove, so this event says "as the user's mouse goes over the button, execute this code." It's as simple as that. The code we're executing is to set the value of our inputs with IDs of x and y to equal *the x and y values of the event*. Here's the magic of the event object: it carries with it a *lot* of information. Go ahead and add a line inside this function as console.log(e) and take a look at what's logged, as shown in the following screenshot:

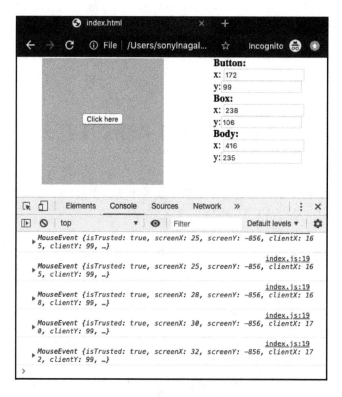

Figure 7.4 – Logging events

As expected, every time your mouse moves over **Click Here**, the event fires and **MouseEvent** is logged out. Open one of those events. You'll see something similar to the following:

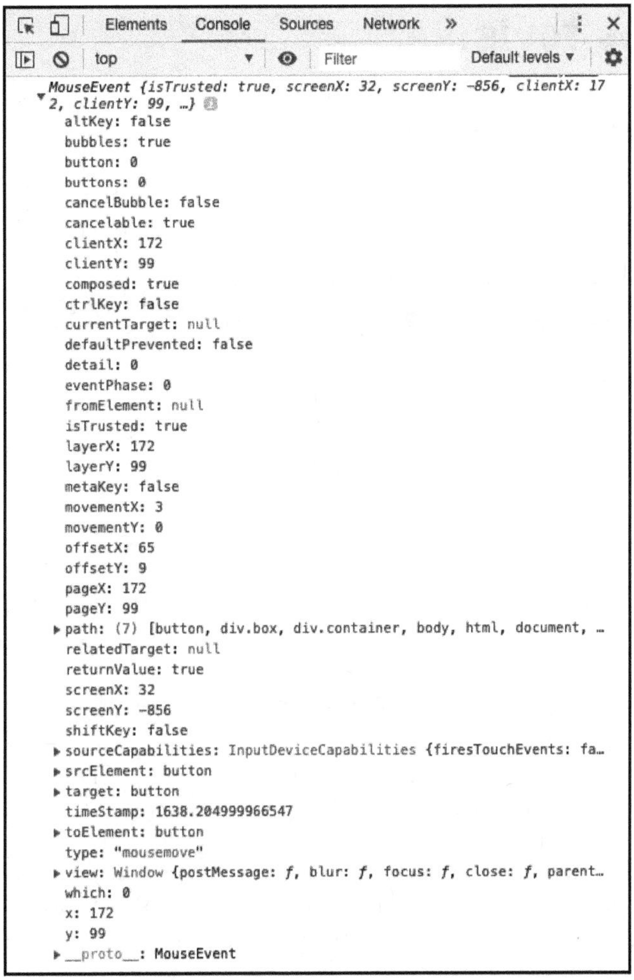

```
⌖  ⬚  │  Elements   Console   Sources   Network   »              ⋮  ✕
▶  ⊘  │  top           ▼  ◉  │  Filter              Default levels ▼  ⚙
   MouseEvent {isTrusted: true, screenX: 32, screenY: -856, clientX: 17
 ▼ 2, clientY: 99, …} ⓘ
     altKey: false
     bubbles: true
     button: 0
     buttons: 0
     cancelBubble: false
     cancelable: true
     clientX: 172
     clientY: 99
     composed: true
     ctrlKey: false
     currentTarget: null
     defaultPrevented: false
     detail: 0
     eventPhase: 0
     fromElement: null
     isTrusted: true
     layerX: 172
     layerY: 99
     metaKey: false
     movementX: 3
     movementY: 0
     offsetX: 65
     offsetY: 9
     pageX: 172
     pageY: 99
   ▶ path: (7) [button, div.box, div.container, body, html, document, …
     relatedTarget: null
     returnValue: true
     screenX: 32
     screenY: -856
     shiftKey: false
   ▶ sourceCapabilities: InputDeviceCapabilities {firesTouchEvents: fa…
   ▶ srcElement: button
   ▶ target: button
     timeStamp: 1638.204999966547
   ▶ toElement: button
     type: "mousemove"
   ▶ view: Window {postMessage: ƒ, blur: ƒ, focus: ƒ, close: ƒ, parent…
     which: 0
     x: 172
     y: 99
   ▶ __proto__: MouseEvent
```

Figure 7.5 – MouseEvent

Here, we see plenty of information about the event, including (as expected) the X and Y coordinates of our mouse at that time. Many of these properties will be useful, but one in particular to note is `target`. The target of an event is the node upon which we placed our event listener. From the `target` property, we can get its ID, which is useful if we have one event handler for multiple nodes.

Do you remember our sticky note program from `Chapter 6`, *The Document Object Model (DOM)*? Let's augment it now.

Sticky notes revisited

Let's take a closer look at our sticky note program from Chapter 6, *The Document Object Model (DOM)*—https://github.com/PacktPublishing/Hands-on-JavaScript-for-Python-Developers/tree/master/chapter-7/stickies/starter-code—and include the ability to create a modal window with information about the sticky when clicked on and the ability to delete that sticky, as shown in the following screenshot:

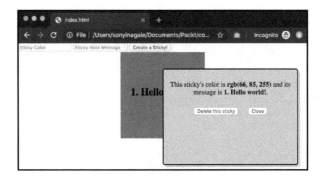

Figure 7.6 – New and improved sticky note creator

To successfully code this, you'll need to use a new DOM manipulation method: .remove(). Take a peek at https://developer.mozilla.org/en-US/docs/Web/API/ChildNode/remove for the documentation. You may also want to take a look at the CSS property of visibility to show and hide the modal window.

Just for fun, I also included a small JavaScript library to use a color picker for the sticky color field as an example of how simple it can be to include third-party code. You don't need to do anything with the jscolor.js script; it will work automatically.

Sticky notes – solution 1

Did you arrive at something similar to the following code?

```
01: const container = document.querySelector('.container') // set
    .container to a variable so we don't need to find it every time
     we click
02: let noteCount = 1 // inital value
03: const messageBox = document.querySelector('#messageBox')
04:
05: // access our button and assign a click handler
06: document.querySelector('.box-creator-button').addEventListener(
```

```
      'click', () => {
07:     // create our DOM element
08:     const stickyNote = document.createElement('div')
09:
10:     // set our class name
11:     stickyNote.className = 'box'
12:
13:     // get our other DOM elements
14:     const stickyMessage = document.querySelector('.box-color-note')
15:     const stickyColor = document.querySelector('.box-color-input')
16:
17:     // get our variables
18:     const message = stickyMessage.value
19:     const color = stickyColor.style.backgroundColor
20:
21:     // blank out the input fields
22:     stickyMessage.value = stickyColor.value = ''
23:     stickyColor.style.backgroundColor = '#fff'
24:
25:     // define the attributes
26:     stickyNote.innerHTML = `${noteCount++}. ${message}`
27:     stickyNote.style.backgroundColor = color
28:
29:     stickyNote.addEventListener('click', (e) => {
30:       document.querySelector('#color').innerHTML =
                  e.target.style.backgroundColor
31:       document.querySelector('#message').innerHTML = e.target.innerHTML
32:
33:       messageBox.style.visibility = 'visible'
34:
35:       document.querySelector('#delete').addEventListener('click', (event)
      => {
36:         messageBox.style.visibility = 'hidden'
37:         e.target.remove()
38:       })
39:     })
40:
41:     // add the sticky
42:     container.appendChild(stickyNote)
43:   })
44:
45:   document.querySelector('#close').addEventListener('click', (e) => {
46:     messageBox.style.visibility = 'hidden'
47:   })
```

You can find this code file on GitHub at `https://github.com/ PacktPublishing/Hands-on-JavaScript-for-Python-Developers/tree/ master/chapter-7/stickies/solution-code-1`.

There are a few interesting pieces here, such as our sticky click handler starting on line 29. Most of this should look familiar, with the addition of a few new friends. First, the click handler is using the event's target property to set text in our message box with attributes of the target. We don't have to search through the DOM in order to find our properties. In fact, doing so would be an expensive and wasteful operation when we have the information passed to us already with the event object. Line 33 modifies the CSS of the modal window to display it, and on line 37, we remove the sticky when the delete button of the modal is clicked.

This works pretty well! However, there's another feature of events that we can use to make our code more efficient, thanks to the event life cycle: *event delegation*.

Sticky notes – solution 2 – event delegation

The principle of **event delegation** is to register an event listener on a parent event and let event propagation tell us what element has been clicked on. Remember our diagram of the event life cycle and the layers through which an event travels? We can use this to our advantage. Take a look at line 37, which is shown here:

```
container.addEventListener('click', (e) => {
  if (e.target.className === 'box') {
    document.querySelector('#color').innerHTML =
     e.target.style.backgroundColor
    document.querySelector('#message').innerHTML = e.target.innerHTML
    messageBox.style.visibility = 'visible'
    document.querySelector('#delete').addEventListener('click', (event) => {
      messageBox.style.visibility = 'hidden'
      e.target.remove()
    })
  }
})
```

You can find this code on GitHub at `https://github.com/ PacktPublishing/Hands-on-JavaScript-for-Python-Developers/blob/ master/chapter-7/stickies/solution-code-2/script.js#L37`.

In this code, we've removed the attachment of the click listener from the sticky creation logic and abstracted it to be attached to the whole container. When `container` is clicked, we check to see whether the target has `box` as its class. If so, we execute our logic! This is a more efficient use of event listeners, especially when used on dynamically created elements. There are cases where event delegation will be your best option and times when either will work.

But now we have another problem: every time a sticky is clicked, a new click handler is added to the delete button. That's not very efficient. See whether you can refactor the code to eliminate that problem.

Sticky notes – solution 3

Here's one possible solution:

```
let target = {}

...

container.addEventListener('click', (e) => {
  if (e.target.className === 'box') {
    document.querySelector('#color').innerHTML =
     e.target.style.backgroundColor
    document.querySelector('#message').innerHTML = e.target.innerHTML
    messageBox.style.visibility = 'visible'
    target = e.target
  }
})

document.querySelector('#delete').addEventListener('click', (event) => {
  messageBox.style.visibility = 'hidden'
  target.remove()
})
```

 You can find this solution on GitHub at `https://github.com/PacktPublishing/Hands-on-JavaScript-for-Python-Developers/blob/master/chapter-7/stickies/solution-code-3/script.js`.

While this uses a global variable, it's still more efficient. By encapsulating our whole program in a function or class, we could eliminate the global variable, but that's not important for this concept.

It's now time to look at Ajax and how events tie into the life cycle of a program. Let's do a lab!

Using Ajax and events to populate API data

Let's put it all together. For this lab, we're going to be creating a simplified Pokémon game using PokéAPI: `https://pokeapi.co/`.

Here's what our game will end up being: `https://sleepy-anchorage-53323.herokuapp.com/`. Go ahead and pull up the site and play around with it to see the functionality.

Please resist the temptation to look at the finished JavaScript file (for now).

Here's a screenshot of what you'll see when you access the preceding URL and start playing the game:

Figure 7.7 – Pokémon game

All of the HTML and CSS have been provided for you. You'll be working in the `main.js` file: `https://github.com/PacktPublishing/Hands-on-JavaScript-for-Python-Developers/tree/master/chapter-7/pokeapi/starter-code`.

If you're not familiar with Pokémon, don't worry! The logic behind this game is basic. (If you *are* familiar with the games, forgive the simplified approach.)

Here's what we'll be doing:

1. Query PokéAPI for all the Pokémon available.
2. Populate the select list with the names of the Pokémon and the value of their API URLs, as provided by the API.

3. When this is done, toggle the CSS property to show the player selections.
4. Allow each of the two players to select their Pokémon.
5. Create functionality for each player to use their Pokémon's moves against the other.
6. Decrement the other player's Pokémon hit points based on a random number generated from the maximum power possible.
7. Display the overlay with text stating that it's effective.
8. If the move does not have the power attribute, display the overlay saying it's not effective.
9. When one Pokémon's hit points are 0 or lower, display the overlay that the opponent has fainted.

Let's break down the starter code.

Starter code

Let's take a look at the starter code, piece by piece, as it introduces a new formulation of our JavaScript: classes! If you're familiar with classes in Python or other languages, this ES6 introduction will come as a welcome reminder of the use of JavaScript. Let's begin:

```
class Poke {
    ...
}
```

First of all, when declaring a class in JavaScript ES6, we simply create an object! Now, the details of the object *are* a little different than what we're used to, but many of the principles are the same. To create an instance of the class, we can say `const p = new Poke()` after finishing the class code.

After that, there is some syntactic sugar with classes, such as constructors, getters, and setters. Feel free to research classes in JavaScript, as it'll help you with the overall goal.

I've given you the starter to a constructor, which is executed when you create an instance of a class:

```
constructor() {
    /**
     * Use the constructor as you would in other languages: Set up your
       instance variables and globals
     */
}
```

What might you want in your constructor? Maybe you want references to often-used DOM elements or event handlers? Then, of course, the question arises: how do we *reference* the variables we've created?

The answer is `this`. When using a variable global to the class, you can preface it with `this.<variableName>` and it will be available to all methods. Here's the great part: it's not a pure global variable to our whole page, but just to our class! If you recall a few of the previous code examples, we didn't handle that piece; here's one way to do so:

```
choosePokemon(url, parent) {
  ...
const moves = data.moves.filter((move) => {
  const mymoves = move.version_group_details.filter((level) => {
    return level.level_learned_at === 1
  })
  return mymoves.length > 0
})
}
```

Since each Pokémon has multiple moves that it learns at different points in the game, this is the logic for finding the moves available at the beginning of play. You won't have to modify it, but take a look at the `.filter()` method of arrays. We didn't cover it before, but it's a useful method to know. MDN is a good resource: https://developer.mozilla.org/en-US/docs/Web/JavaScript/Reference/Global_Objects/Array/filter.

The next part of the code we're interested in is the **setter**:

```
set hp(event) {
  ...
  if (event.hp) {
    this[event.player].hp = event.hp
  }

  if (event.damage) {
    this[event.player].hp -= event.damage
  }
  const e = new CustomEvent("hp", {
    detail: {
      player: event.player,
      hp: this[event.player].hp
    }
  })
  document.dispatchEvent(e)
}
```

A **setter** is a class method that handles setting or changing a member variable. Usually used with a **getter**, the concept allows us to abstract the logic of manipulation needed when changing (or retrieving) a variable. In this case, we're using some game logic to see how to treat the hit points. But then we get into a new and wonderful idea: custom events.

Custom events

With the `new CustomEvent()` directive, we can create a new, named event to use in our program. Sometimes, user interactions or page behavior don't fully handle what we need. Custom events can help with that need. Notice in the preceding code that the `detail` object contains data to be passed with the event, and we use `document.dispatchEvent()` to send it to the event stream. Creating an event listener for a custom event is the same as doing so with a built-in event: use `.addEventListener()`. We'll want to use the `doMove()` function.

Solution code

How did you do with your attempt? You can take a look at one possible way of solving the lab here: `https://github.com/PacktPublishing/Hands-on-JavaScript-for-Python-Developers/tree/master/chapter-7/pokeapi/solution-code`.

Remember, there are multiple ways to solve a programming problem, so if your solution doesn't match the provided approach, it's OK! The main idea is to solve the problem.

Handling asynchronicity

As we can see when using APIs, the asynchronous nature of Ajax calls for a couple of creative approaches. In our Pokémon game, we used a loading spinner while calls were completing; this is an approach you've seen all around the modern web. Let's take a look at one example from the game:

```
toggleLoader() {
  /**
    * As this is visual logic, here's the complete code for this function
    */
  if (this.loader.style.visibility === 'visible' ||
  this.loader.style.visibility === '') {
    this.loader.style.visibility = 'hidden'
  } else {
```

```
            this.loader.style.visibility = 'visible'
        }
    }
```

All *this* part of the code is doing is toggling the visibility of a layer that contains a spinning image. This is all in the CSS (as it's not technically an image, but rather a CSS animation). Let's look at how it's used:

```
getPokemon() {
    fetch('https://pokeapi.co/api/v2/pokemon?limit=1000')
        .then((response) => {
          return response.json()
        })
        .then((data) => {
          const pokeSelector = document.querySelector('.pokeSelector.main')

          data.results.forEach((poke) => {
            const option = document.createElement('option')
            option.value = poke.url
            option.innerHTML = poke.name
            pokeSelector.appendChild(option)
          })

          const selector = pokeSelector.cloneNode(true)
          document.querySelector('.pokeSelector.clone').replaceWith(selector)

          this.toggleLoader()

          document.querySelector('#Player1').style.visibility = 'visible'
          document.querySelector('#Player2').style.visibility = 'visible'
        })
    }
```

Here, we see that in our asynchronous Promise calls with `.then()`, we're toggling the loader when everything is complete! It's a nice little bundle. If you'd like a refresher on how to use `fetch` and Ajax calls in general, take a look back at Chapter 4, *Data and Your Friend, JSON*, in the *API calls from the frontend – Ajax* section.

When dealing with the inherently asynchronous nature of Ajax calls, it's important to keep in mind the fact that we don't know exactly when a call will return with its data—or even *if* it will return at all! We can make our code better with **error handling**.

Error handling

Take a look at this code:

```
fetch('/profile')
  .then(data => {
    if (data.status === 200) {
      return data.json()
    }
    throw new Error("Unable to get Profile.")
  })
  .then(json => {
    console.log(json)
  })
  .catch(error => {
    alert(error)
  })
```

We have some of the usual suspects here: a `fetch` call and `.then()` handling our results. Now, take a look at `new Error()` and `.catch()`. Just like in most languages, JavaScript has a way to explicitly throw errors, and `.catch()` at the end of our `fetch` chain will then present the error to the user in an alert box. It's always best practice to include error handling in your Ajax calls in case the service you're calling doesn't respond, doesn't respond in time, or sends back an error. We'll discuss errors a bit more in `Chapter 9`, *Deciphering Error Messages and Performance Leaks*.

Star Wars API exploration Lab

Let's get our hands dirty with some more Ajax calls. We'll be using the popular **Star Wars API (SWAPI)**: `https://swapi.dev/` . Take a few minutes to familiarize yourself with the documentation and how the API works.

Here's what we'll be building:

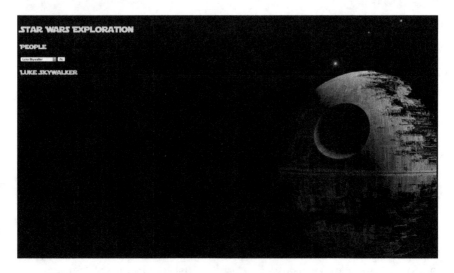

Figure 7.8 – Star Wars exploration

You can experiment with the functionality at `https://packtpublishing.github.io/Hands-on-JavaScript-for-Python-Developers/chapter-7/swapi/solution-code/`. Try to resist the temptation to peruse the solution code until after you've tried your hand at recreating the functionality.

Here's what our code should do:

1. Display a loader on page load. This loader is provided for you as a CSS animation.
2. Call the `/people` SWAPI endpoint to retrieve all people in the API. *Hint: you will need to call SWAPI more than once to get all of the people.*
3. Populate the select list with the names of the people and hide the loader.
4. When **Go** is clicked, make another call to SWAPI to retrieve details about the person chosen and display them (at least the name).

Our approach will be to first populate the list and then prepare for user action, in order to explore both synchronously chained events and asynchronous action dependent on user input.

The starter HTML and CSS shouldn't need to be altered, and our starter JavaScript file is nearly empty! Are you ready for the challenge? Good luck!

A solution

If you look at the solution code, you'll find one way of creating this functionality. Let's break it down.

Just like in our Pokémon game, we'll use a class. Its constructor will store a few various pieces of information, as well as add an event listener to the **Go** button:

```
class SWAPI {
  constructor() {
    this.loader = document.querySelector('#loader')
    this.people = []

    document.querySelector('.go').addEventListener('click', (e) => {
      this.getPerson(document.querySelector('#peopleSelector').value)
    })
  }
}
```

Next, as we know we'll be making multiple calls to SWAPI, we can make ourselves a helper function to facilitate this work. It may take four arguments: the SWAPI API URL, an array of previous results (useful if we're paginating!), and Promise-like `resolve` and `reject` parameters:

```
fetchThis(url, arr, resolve, reject) {
  fetch(url)
    .then((response) => {
      return response.json()
    })
    .then((data) => {
      arr = [...arr, ...data.results]
```

This last line may be new. ... is the spread operator and it expands an array into each of its parts. With this ES6 feature, we don't need to iterate over an array in order to join it to another or do any other reassignment magic. We can simply explode the results and join them with the existing results:

```
      if (data.next !== null) {
        this.fetchThis(data.next, arr, resolve, reject)
      } else {
        resolve(arr)
      }
```

In many APIs, if the dataset is large, only limited results will be returned, with a link for the next and previous pages of data. SWAPI's nomenclature specifies that `.next` is the property to look for if there's another page. Otherwise, we can return our results in our `resolve` function:

```
    })
    .catch((err) => {
      console.log(err)
    })
```

Don't forget error handling!

```
  }

getPeople() {
    new Promise((resolve, reject) => {
        this.fetchThis('https://swapi.dev/api/people', this.people,
        resolve, reject)
    })
    .then((response) => {
      this.people = response
      const peopleSelector = document.querySelector('#peopleSelector')

      this.people.forEach((person) => {
        const option = document.createElement('option')
        option.value = person.url
        option.innerHTML = person.name
        peopleSelector.appendChild(option)
      })
      this.toggleLoader()
      document.querySelector('#people').style.visibility = 'visible'
    })
    .catch((err) => {
      console.log(err)
    })
  }
```

Try to read through `getPeople()` in its entirety to get a sense of what it does. Some of it is simple manipulation, but `new Promise()` is the core of this function. Instead of hardcoding page numbers to iterate over our API's list of people, we're creating a new Promise that uses our `fetchThis` function:

```
getPerson(url) {
    this.toggleLoader()
    fetch(url)
      .then((response) => {
        return response.json()
```

```
  })
  .then((json) => {
    document.querySelector('#person').style.visibility = 'visible'
    document.querySelector('#person h2').innerHTML = json.name
    this.toggleLoader()
  })
  .catch((err) => {
    console.log(err)
  })
}
```

In theory, we can use the same `fetchThis` function to get an individual person as well, once the button is clicked, but just for the sake of our example, this solution handles it all in one piece:

```
toggleLoader() {
  if (this.loader.style.visibility === 'visible' ||
  this.loader.style.visibility === '') {
    this.loader.style.visibility = 'hidden'
  } else {
    this.loader.style.visibility = 'visible'
  }
}
}
```

Then, all that's needed is to instantiate our class!

```
const s = new SWAPI().getPeople()
```

At this point, our program is complete and will run! Visit the page and you'll see our fully operational page. Emperor Palpatine thanks you for your help eliminating the rebel scum. We have seen the power of classes, event-based programming, and our ability to use events to our benefits.

Summary

We've learned about events, their life cycle, and how event-driven design works. An **event** is triggered by an action by the user (or programmatically based on the program logic) and enters its **life cycle**. In the event life cycle, our program can pick up on many pieces of information carried by the **event object** itself, such as the mouse position or target DOM node.

By understanding how Ajax works with events, you're well on your way toward becoming a fully fledged JavaScript developer. **Ajax** is incredibly important as its the conduit between JavaScript and external APIs. Since JavaScript is stateless and client-side JavaScript has no concept of sessions, it's important for Ajax calls to be **asynchronous** in nature; hence the introduction of tools such as `fetch`.

Congratulations! We've covered a lot of very dense material. Next up are frameworks and libraries in JavaScript.

Questions

Answer the following questions to gauge your understanding of events:

1. Which of these is the second phase of the event life cycle?
    1. Capturing
    2. Targeting
    3. Bubbling

2. The event object provides us with which of the following? – Select all that apply:
    1. The type of event triggered
    2. The target DOM node, if applicable
    3. The mouse coordinates, if applicable
    4. The parent DOM node, if applicable

Look at this code:

```
container.addEventListener('click', (e) => {
  if (e.target.className === 'box') {
    document.querySelector('#color').innerHTML =
     e.target.style.backgroundColor
    document.querySelector('#message').innerHTML = e.target.innerHTML
    messageBox.style.visibility = 'visible'
    document.querySelector('#delete').addEventListener('click', (event) =>
{
      messageBox.style.visibility = 'hidden'
      e.target.remove()
    })
  }
})
```

3. Which JavaScript features ae used in the preceding code? Select all that apply:
    1. DOM manipulation
    2. Event delegation
    3. Event registration
    4. Style changes

4. What will happen when the container is clicked?
    1. `box` will be visible.
    2. `#color` will be red.
    3. Both 1 and 2.
    4. There is not enough context.

5. In which phase of the event life cycle do we typically take action?
    1. Targeting
    2. Capturing
    3. Bubbling

Further reading

- *JavaScript: Understanding DOM Event Life Cycle*: `https://medium.com/prod-io/javascript-understanding-dom-event-life-cycle-49e1cf62b2ea`
- w3schools.com – JavaScript events: `https://www.w3schools.com/js/js_events.asp`
- MDN – event reference: `https://developer.mozilla.org/en-US/docs/Web/Events`

Working with Frameworks and Libraries

8

Very few languages exist in a self-contained, monolithic ivory tower. Almost always, especially with any modern language, third-party code is used in programs for added functionality. Using third-party code, such as libraries and frameworks, is also an integral part of working with JavaScript. Let's examine a few of the more popular open source tools in our toolkit.

The following topics will be covered in this chapter:

- jQuery
- Angular
- React and React Native
- Vue.js

Technical requirements

Be prepared to work with the code provided in the `Chapter-8` directory of the repository: `https://github.com/PacktPublishing/Hands-on-JavaScript-for-Python-Developers/tree/master/chapter-8`. As we'll be working with command-line tools, also have your Terminal or command-line shell available. We'll need a modern browser and a local code editor.

jQuery

One of the principal reasons for creating or using a JavaScript library is to ease repetitive or complex tasks. After all, you can't fundamentally *change* a language with a plugin or library—all you can do is augment or alter the existing functionality.

As we discussed in Chapter 1, *The Entrance of JavaScript into Mainstream Programming*, JavaScript's early history was a bit of a Wild West scenario. The browser wars were in full effect, features were not standardized, and even making an Ajax call required two different sets of code: one for Internet Explorer and one for the other browsers.

Enter jQuery in 2006, created by John Resign.

The lack of standardization across browsers was the impetus for creating jQuery. From DOM manipulation to Ajax calls, the syntax and structure of jQuery are a "write once, use in all browsers" paradigm. With the development of ES6 and beyond, JavaScript *is* getting more standardized. However, there's over a decade of jQuery code out there that the majority of JavaScript-heavy websites use. Because of these legacy applications, it still enjoys quite a bit of popularity, so it's important to our discussion. It's also open source, so there are no licensing fees associated with using it.

Advantages of jQuery

Consider the following examples, which do the same thing:

- **JavaScript ES6**: `document.querySelector("#main").classList.add("red")`

- **jQuery**: `$("#main").addClass("red");`

As you can see, jQuery construction is much shorter. Great! Concise code is usually a good thing. So, let's break down this example:

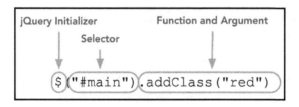

Figure 8.1 – jQuery syntax

1. We start almost all jQuery statements with `$`. This is a convention that's been used in many libraries, and actually, it's possible to override the dollar notation and use anything you'd like, so you may see examples that start with `jQuery`.

2. Our selector is a CSS selector, just as we use with `document.querySelector()`. A convention is to preface DOM nodes that you've selected via jQuery with dollar notation if you store them for later use. So, if we were to store `#main` as a variable, it might look like: `const $main = $("#main")`.

3. jQuery has its own list of functions that are usually legible shorthands for internal functionality.

One interesting fact about jQuery: you can mix jQuery with vanilla JavaScript (which is not using any frameworks or libraries). In fact, the term "vanilla JavaScript" is a popular way of referring to non-jQuery code.

Additionally, some frontend libraries, such as Bootstrap, prior to Bootstrap 5, are built with jQuery, so understanding its usage can help you understand other libraries and frameworks. It's not a *bad* thing, but important to be aware of as you navigate the brave new world of frontend development.

Disadvantages of jQuery

Using jQuery, as with any library, requires an additional download on the client side. As of the time of writing, jQuery version 3.4.1's minified version clocks in at 88 KB. Now, while that's pretty negligible and will be cached by the browser, keep in mind that this must be executed and loaded on every single page, so it's not just the download size but also the execution time that should be considered. Wes Bos also has some great information about scoping in ES6 versus jQuery: `https://wesbos.com/javascript-arrow-functions/`.

Also, while not true in all cases, much of jQuery's usage exists to standardize ES5, so the majority of code you'll see online and in examples will be ES5.

Examples of jQuery

Let's compare our original Star Wars exploration from `Chapter 7`, *Events, Event-Driven Design, and APIs* (`https://github.com/PacktPublishing/Hands-on-JavaScript-for-Python-Developers/tree/master/chapter-8/swapi`) with a jQuery version (`https://github.com/PacktPublishing/Hands-on-JavaScript-for-Python-Developers/tree/master/chapter-8/swapi-jQuery`).

Now, I'll grant that this is not the most elegant piece of jQuery, but there are reasons for why that is so. Let's break this down.

First up—HTML:

ES6	jQuery
No change	Addition of `<script src="https://code.jquery.com/jquery-3.4.1.min.js"></script>`

As we discussed, adding in a JavaScript library or framework inherently needs another file download and/or overhead processing time from local files. Usually, the size is negligible, so in this case, the only relevant factor is that we're adding a line of HTML to load the jQuery file from a global content delivery network.

There will be no change to the CSS, as expected. So let's dive into the JavaScript:

ES6	jQuery

<pre>class SWAPI {
  constructor() {
    ...
  }
}</pre>
	<pre>var swapi;
	
	$(document).ready(function() {
	  swapi = new SWAPI;
	});</pre>

OK, now we're seeing some major differences. As mentioned, this isn't necessarily the most ideal jQuery program, but I think it gets the point across. To start with, while jQuery and ES6 are compatible, most jQuery is used where ES6 isn't available—or the code hasn't been upgraded yet to ES6. One of the first things you'll notice about most jQuery code is that it uses semicolons at the end of lines and uses `var` instead of `let` or `const`. This isn't unique to jQuery; rather, they are ES5 conventions.

Instead of using classes, ES5 usually uses manipulation of an object prototype, as follows:

```
SWAPI.prototype.constructor = function() {
  this.$loader = $('#loader');
  this.people = [];
};
```

Classes are arguably cleaner ways to do work because they are more self-contained and explicit in their methods and usage. However, this convention wasn't around when jQuery was popular, so we'll use ES5 prototypal inheritance.

Let's now take a side-by-side look at how making Ajax calls with ES6 and jQuery differ:

ES6	jQuery
| ```
fetch(url)
  .then((response) => {
    return response.json()
  })
  .then((json) => {

    ...
  })
``` | ```
$.get(url)
  .done(function(data) {

    ...
  };
``` |

Here's a great example of the why of jQuery and how its creation contributed to some simplifications in ES6. In ES5, making an Ajax request required two different methods of doing so—one for Internet Explorer and one for other browsers—as the requesting methods weren't standardized. jQuery helped by doing that browser detection and code-switching *under the hood* so that a developer only needed to write one statement. With `fetch`, however, this is no longer needed. However, we do see that the jQuery code is a little shorter because we don't have the first `.then` function to return the JSON from the request. Is this a design flaw or feature? It's actually the latter, as APIs could return many different kinds of responses. The `fetch` method does some conversion for you under the hood, whereas jQuery expects you to pretty much know what your data is and how to work with it.

W3Schools has great examples and reference material on jQuery: `https://www.w3schools.com/jquery/`.

If you review the rest of the code in the jQuery version of the code, you'll find many other interesting examples of differences, but for now—onward from jQuery! Let's take a look at a fully fledged *web framework*: Angular.

Angular

Angular was created by Google as *AngularJS*. In 2016, it was rewritten as version 2, separating it from AngularJS. It's open source and a framework, versus a library, which now raises the question: what's the difference between a **framework** and a **library**?

A *library* is a toolkit to more easily write your code, for varying purposes. To use an architectural analogy, a library is like a set of bricks that can be used to assemble a house. Conversely, a *framework* is more aligned with the blueprints used to design the house. It may use some of the same bricks—or it might not! One of the main differences is that libraries, in general, allow you to write your code as you would like to write it without the library dictating opinions on how to structure your code. Frameworks, on the other hand, are more opinionated and ask you to structure your code in the best practices of *that* framework. It's a nebulous (and sometimes overloaded) term, so there's an understandable amount of debate on what's a library and what's a framework. Just search *Stack Overflow* and you'll find competing definitions. A good simplified statement is that a **framework** can be a collection of technologies with a specified use pattern, whereas a **library** is more likely to be one technology that helps manipulate your data.

Let's consider this diagram:

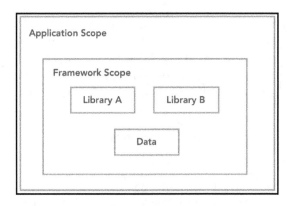

Figure 8.2 – Framework composition

As we can see, a framework can, in fact, be comprised of multiple libraries. The design pattern of the framework usually dictates how and when these libraries are used.

Angular uses *TypeScript*, which is an open source programming language. Originally developed by Microsoft, it's JavaScript with some additional functionality that is—or can be—appealing to some developers. Even though TypeScript is classified as its own language, it's a superset of JavaScript, so it transpiles down to normal JavaScript and thus doesn't require any additional work to run in a browser, aside from executing the Angular build process.

Advantages of Angular

Angular, like most frameworks, is opinionated about your file structure and code syntax (especially with TypeScript in the mix). This may sound like a disadvantage, but it's actually really important when working in a team: you have been seeded with existing file structures regarding how to work with your code, and this is a *good* thing.

Angular also does not exist in isolation. It's part of a **technology stack**, which means that it's a soup-to-nuts solution encompassing the frontend to a database. You may have encountered the term **MEAN** stack: **MongoDB, Express, Angular, and Node.js**. While you can use Angular outside of this stack, it provides an easy-to-setup ecosystem for development that is well understood by others.

If you're not familiar with the **Model-View-Controller** (**MVC**) paradigm, now is a good time to get acquainted with it. Many technology stacks across multiple languages utilize this paradigm to separate concerns in your code base. For example, the **model** in your program works with the data acquisition and manipulation from your data source(s) (such as a database and/or APIs), while the **controller** manages the interactions between the model, data sources, and the **view** layer. The **view** mostly controls the visual display of the information in a full-stack environment. There is debate within the full-stack MVC community, as far as approaches go, between the so-called "fat model, skinny controller" approaches, and the converse. It's not important right now to go into that distinction, but you'll see that debate in the community.

Speaking of community, there's the fact that Angular developers have formed an ad hoc network of people helping each other out. The discussions alone are valuable and help you navigate the landscape.

There are a few other advantages to Angular, such as two-way data binding (making sure the model and view talk to each other) and specialized directives bound to HTML elements, but those are nuances that aren't important to discuss right now.

Disadvantages of Angular

The main con of Angular is its steep learning curve. Along with the discrepancy between the original AngularJS and the more modern Angular iterations, Angular is, unfortunately, suffering from decreasing popularity among developers. Additionally, it is *quite* verbose and complex. According to some Angular developers, tasks such as working with third-party libraries can be repetitive.

The use of TypeScript instead of standard ES6 is also a point of concern. While TypeScript is useful, it adds to the learning curve of using Angular. That being said, Angular is indeed very versatile.

Examples of Angular

Let's build a small "Hello World" application with Angular. We'll need a few tools to start our work, such as `npm`. Refer to `Chapter 2`, *Can We Use JavaScript Server-Side? Sure!*, for installation of `npm` and its friends. If you'd prefer, you can also follow along with the code provided at `https://github.com/PacktPublishing/Hands-on-JavaScript-for-Python-Developers/tree/master/chapter-8/angular-example`.

Here are our steps:

1. Begin by installing the Angular CLI: `npm install -g @angular-cli`.
2. Create a new example project with `ng new example`. Accept the defaults for this installation by pressing *Enter* at the prompts.
3. Go into the directory that was just created: `cd example`.
4. Begin the server: `ng serve --open`.

At this point, your web browser should open this page at `http://localhost:4200/`:

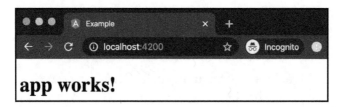

Figure 8.3 – Example start page

OK, great. This looks like a simple enough page for us to work with. Here's the file structure that our CLI created:

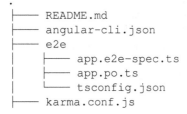

```
.
├── README.md
├── angular-cli.json
├── e2e
│   ├── app.e2e-spec.ts
│   ├── app.po.ts
│   └── tsconfig.json
├── karma.conf.js
```

```
├──── package-lock.json
├──── package.json
├──── protractor.conf.js
├──── src
│     ├──── app
│     │     ├──── app.component.css
│     │     ├──── app.component.html
│     │     ├──── app.component.spec.ts
│     │     ├──── app.component.ts
│     │     └──── app.module.ts
│     ├──── assets
│     ├──── environments
│     │     ├──── environment.prod.ts
│     │     └──── environment.ts
│     ├──── favicon.ico
│     ├──── index.html
│     ├──── main.ts
│     ├──── polyfills.ts
│     ├──── styles.css
│     ├──── test.ts
│     └──── tsconfig.json
└──── tslint.json
```

Let's look at the code that was generated. Open `src/index.html`. Here's what you'll see:

```html
<!doctype html>
<html lang="en">
<head>
  <meta charset="utf-8">
  <title>Example</title>
  <base href="/">

  <meta name="viewport" content="width=device-width, initial-scale=1">
  <link rel="icon" type="image/x-icon" href="favicon.ico">
</head>
<body>
  <app-root></app-root>
</body>
</html>
```

That's it! You see, this is just a template for Angular to create the page we were just looking at, and then the JavaScript does the rest. If you view the source of the page in the browser, you'll see something very similar, but just with a few script calls. All JavaScript for a **single-page app** (**SPA**) is downloaded at once or possibly chunked into blocks intended to be used in harmony.

Single-page applications

It's worth discussing exactly what a SPA is. We've touched on this topic before, but let's now take a look at why Angular (and React and Vue, which we'll get to shortly) are popular and compelling to use. Think of a standard, HTML-based website. It probably has a consistent header, footer, and styling. However, a standard website needs to download (or serve from local cache) these assets every time you navigate to a different page (not to mention retrieving the HTML and rerendering it). A SPA eliminates this redundancy by packaging together all of the relevant data into one unified package that's transmitted to the browser. The browser then parses the JavaScript and renders it. The result is a fast, smooth experience that essentially eliminates page load time lag. You've used these already. If you use Gmail or most of the modern online email systems, you may have noticed that the page load time is negligible, or, at worst, has a small loading icon. The load time and ostensibly wasteful redownload of resources and content is one problem that SPAs are designed to handle.

Now that we've discussed how a SPA can help increase our efficiency, let's take a look at the JavaScript behind our Angular example.

The JavaScript

First of all, let's open `src/app/app.component.html` and look at line 2: `{{ title }}!`.

Hm, what are these curly braces? If you're familiar with other templating languages, you may recognize this as a template token that's intended to be replaced by our rendering language before being rendered. So, what is the method to replace it?

Let's now look at `src/app/app.component.ts`:

```
import { Component } from '@angular/core';

@Component({
  selector: 'app-root',
  templateUrl: './app.component.html',
  styleUrls: ['./app.component.css']
})
export class AppComponent {
  title = 'app works!';
}
```

We can see that the template is referencing `app.component.html` and our `AppComponent` class is specifying `title` as `app works!`. That's exactly what we saw in our browser. Welcome to the power of a templating system!

For now, we won't get into the SPA feature of Angular, but check out the Angular tutorial at `https://angular.io/tutorial` for more details.

Now, let's continue on our tour with React.

React and React Native

Originally created by Jordan Walke at Facebook in 2013, React has quickly evolved into one of the leading user interface libraries currently in use. In contrast with Angular, React does not seek to be a complete framework, but rather focuses on specific parts of the web workflow. Since web pages are inherently *stateless* (that is, no real information is transferred from page to page), SPAs aim to store certain pieces of state in JavaScript memory, enabling subsequent views to be populated with data. React is a prime example of how this type of architecture works while still not encompassing the entire framework paradigm. In MVC terminology, React deals with the view layer.

Advantages of React

Since React *itself* only deals with views, it relies on other libraries to round out its feature set, such as React Router and Hooks. That is, the base architecture of React is designed to be modular and have add-ons used to do other parts of the workflow. At the moment, it's not important to know about React Router, Hooks, or Redux, but just be aware that React is only one part of the puzzle for a complete website.

So, why is this an advantage? Unlike some other JavaScript tools, such as Angular, React doesn't try to reinvent the wheel with its own rules and regulations or language structures. It feels like you're coding in basic JavaScript because, for the most part, you are!

Another advantage of React is how it deals with components and templates. Components are simply reusable pieces of code that can be used in multiple places in your program with different data to populate the view. React also has a great step-by-step tutorial at `https://reactjs.org/tutorial/tutorial.html`. We'll dissect this in the *Examples of React* section. For now, of course, we need to discuss the disadvantages.

Disadvantages of React

To be honest, the learning curve for React (and especially its newer sister technologies, such as Redux and Hooks, which simplify state-based management) is steep. However, by the community, that's not even considered a major disadvantage, because the same is true with almost all libraries and frameworks. A major disadvantage, however, is its rapid pace of development. Now, you may be thinking: *but a continually evolving technology is good*! That is good thinking, but in practice, it can be a bit daunting, especially when dealing with breaking changes.

Another turn-off for some developers is the mixing of HTML and JavaScript inside JavaScript. It uses a syntax extension that allows adding HTML within your JavaScript that is called JSX. For purists, the mixing of presentation layer code into logic structures may seem foreign and an architectural anti-pattern. There is, again, a learning curve to JSX.

It's time to take a look at a classic React example application: Tic Tac Toe.

Example of React

You can follow along a step-by-step tutorial to build this application at `https://reactjs.org/tutorial/tutorial.html`, and for ease of use, you can use this GitHub directory—`https://github.com/PacktPublishing/Hands-on-JavaScript-for-Python-Developers/tree/master/chapter-8/react-tic-tac-toe`—for the completed example:

1. Clone the repository and `cd` into the `react-tic-tac-toe` directory.
2. Execute `yarn start`.

   Don't be surprised by the new `yarn` command. It's a different package manager that is similar to `npm`.

3. When `yarn start` is complete, it will provide you with a URL similar to `http://localhost:3000/`. Open it in your browser. You should see this:

Figure 8.4 – React Tic Tac Toe, starting

If you're not familiar with the game Tic Tac Toe, the logic is simple. Two players alternate marking an **X** or an **O** in a 3-by-3 grid until one player has three of their marks in a row, whether horizontally, vertically, or diagonally.

Let's play! If you click the boxes, here's what you may arrive at:

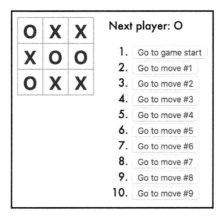

Figure 8.5 – React Tic Tac Toe, possible end state

Notice that the example also maintains a state history with the buttons on the right side of the screen. You can rewind play into any of those states by clicking the buttons. This is an example of how React uses **state** in order to maintain a sense of continuity across parts of the application.

Components

To illustrate the concept of reusable components, consider the code for the top row of the Tic Tac Toe grid. Take a peek at `src/index.js`.

You should see this starting on line 27:

```
<div className="board-row">
  {this.renderSquare(0)}
  {this.renderSquare(1)}
  {this.renderSquare(2)}
</div>
```

`renderSquare` is a fairly simple function that renders JavaScript XML, or **JSX**. As mentioned before, JSX is an extension to JavaScript. It introduces XML-like capabilities in a standard JavaScript file, marrying JavaScript syntax with a set of HTML and XML to construct the components we've been talking about. It's not its own fully fledged templating language per se, but it might, in some ways, actually be more powerful.

Here's `renderSquare`:

```
renderSquare(i) {
  return (
    <Square
      value={this.props.squares[i]}
      onClick={() => this.props.onClick(i)}
    />
  );
}
```

So far, so good…it's fairly standard-looking…except for one thing. What is `Square`? That's not an HTML tag! Here's the power of JSX: we can define our own reusable tags as these wonderful components we've been talking about. Think of them as LEGO® bricks we can use to assemble our own application. From basic building blocks, we can construct a very sophisticated SPA.

Thus, `Square` is simply a function that returns a standard HTML button with a few properties, such as its `onClick` handler. You can see what this handler does later on in the code:

```
function Square(props) {
  return (
    <button className="square" onClick={props.onClick}>
      {props.value}
    </button>
  );
}
```

We've really only scratched the surface of React, but I hope you've gotten a sense of its power. In fact, it's poised to be the dominant frontend framework in the ecosystem. At the time of writing, React has far eclipsed Angular in the number of job openings in the technology world.

React Native

A discussion of React would be incomplete without mentioning React Native. One of the difficult aspects of native mobile application development is, well, native languages. The Android platform utilizes Java, while iOS relies on Swift as a programming language. We won't get into the details of mobile development here (or React Native), but it's important to note that there are major differences between React and React Native. When I first began experimenting with React, I was under the impression that the components were reusable between React and React Native. To an extent, this is *mildly* true, but the differences between the two outweigh the similarities.

Native's main strength comes from the fact that you're not writing in another language; rather, you're still writing JavaScript. With that being said, there are additional complexities to Native, especially when dealing with the native features of a mobile device, such as the camera. To that end, I advise you to consider carefully the use of React Native in your project lifespan and do *not* assume all knowledge transfers from one to the other.

Next, let's discuss a relative newcomer to the JavaScript world: Vue.js.

Vue.js

Another newcomer to the JavaScript framework ecosystem is Vue.js (more commonly referred to simply as Vue). Developed in 2014 by Evan You, it's another open source framework that's designed to provide advanced functionality for SPAs and user interfaces. Evan You felt that there were worthwhile parts of Angular worth keeping, yet room for improvement. It's an admirable goal! Some may say the project succeeded in doing just that, while others find other projects superior. However, the goal of this chapter isn't to pass judgment on any technology but rather to expose you to various extensions of JavaScript that will make your work easier and more adherent to modern standards.

Vue, unlike React, includes *routing, state,* and *build tooling* out of the box. It also has a learning curve, as do many similar technologies, so be sure to give yourself space and time to learn if you choose to explore Vue.

We'll be investigating the base example of Vue from the official guide at `https://vuejs.org/v2/guide/`. If you look at the lesson in the *Declarative Rendering* section, you'll find a Scrimba lesson. Feel free to watch the tutorial or access the code from `https://github.com/PacktPublishing/Hands-on-JavaScript-for-Python-Developers/tree/master/chapter-8/vue-tutorial`, but here are the basics.

The HTML of Vue looks fairly similar to any other framework that uses curly-brace tokens for content replacement:

```html
<html>
    <head>
        <link rel="stylesheet" href="index.css">
        <script src="https://cdn.jsdelivr.net/npm/vue/dist/vue.js"></script>
    </head>
    <body>
        <div id="app">
            {{ message }}
        </div>
        <script src="index.js"></script>
    </body>
</html>
```

It's worth noting that the curly-brace syntax will likely conflict with other templating systems, such as Mustache, but we'll continue using the built-in Vue technologies for now.

Since you have the `{{ message }}` token, let's see what powers it.

If you look at the `index.js` file, you'll find it is very simple:

```javascript
var app = new Vue({
    el: '#app',
    data: {
        message: 'Hello Vue!'
    }
});
```

This basic structure should look familiar: it's an instantiation of a class with an object as a parameter. Note that the data element contains a key of the message with the value `Hello Vue`. This is what is passed to the view layer as `{{ message }}` and thus our app renders our message:

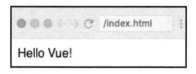

Figure 8.6 – Vue's "Hello World" example

So far, its abilities seem similar to other tools we've explored, so let's dive into the advantages and disadvantages.

Features of Vue.js

As the only competitor to Vue in practice is React, it might be enough to leave this comparison in your hands: `https://vuejs.org/v2/guide/comparison.html`. However, let's break down a couple of points of the comparison with a more objective eye, as even the author(s) of the comparison admitted it was biased toward Vue (as one would expect):

- **Performance**: Ideally, any framework or library adds only a negligible load time or instantiation time to an application, but in practice, this varies. I'm sure we all remember the days of multi-second Ajax or Flash (or even Java servlet!) loaders, but in general, these delays have been mitigated by an asynchronous, step-by-step loading pattern. One of the signature details of a modern web technology should be its unobtrusive and progressive enhancement of the user experience. To that point, Vue does an excellent job of being an additive experience.
- **HTML + JavaScript + CSS**: Vue allows an unprecedented mixing and matching of technologies, whereby it can take standard HTML, CSS, and JavaScript combined with JSX and Vue-specific syntax in order to construct an application. It's a bit of a mixed bag to say whether this is a benefit or a liability, but it is a fact of the technology.
- **Angular ideas**: Unlike React, which rejects almost all Angular conventions, Vue leverages a few learning points from Angular. This could make it a worthwhile framework for someone looking to leave Angular, though the jury is still out on the value/efficacy of this approach.

Now, let's look at an example of Vue.

Example of Vue.js

Let's work with the Vue CLI to create a sample project:

1. Install the CLI with `npm install -g @vue/cli`.
2. Execute `vue create vue-example` in a new directory. For our purposes, you can use the default options by pressing *Enter* at each prompt.
3. Enter the directory: `cd vue-example`.

4. Start the program with `yarn serve`:

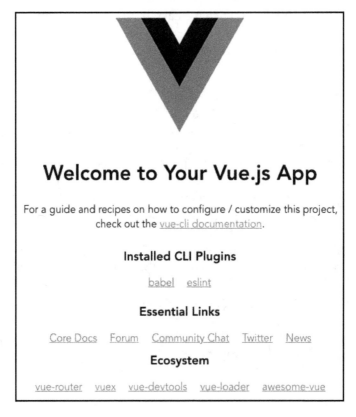

Figure 8.7 – Vue generator home page

Vue's CLI generator created a number of files for us in the `vue-example` directory:

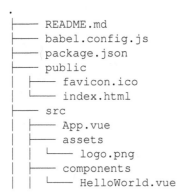

```
.
├── README.md
├── babel.config.js
├── package.json
├── public
│   ├── favicon.ico
│   └── index.html
├── src
│   ├── App.vue
│   ├── assets
│   │   └── logo.png
│   ├── components
│   │   └── HelloWorld.vue
```

```
|    └──── main.js
└──── yarn.lock
```

Let's examine the pieces that it created for us:

1. Open `src/App.vue`. We'll see this in the script block:

   ```
   import HelloWorld from './components/HelloWorld.vue'

   export default {
     name: 'app',
     components: {
       HelloWorld
     }
   }
   ```

   We don't see any of the links we saw in our browser, but the `import` line tells us where the content is.

2. Open `src/components/HelloWorld.vue`. Now, we see the content of our page in a `<template>` node. Feel free to change some of the markup and experiment with the different variables.

And that's Vue in a nutshell! You'll find that after going through Angular and React, the concepts in Vue are a logical progression and will not be difficult to pick up.

Summary

Frontend frameworks are powerful tools, but they're not interchangeable. Each has its pros and cons and your use of them should not only be dictated by what's en vogue at the moment but also by the community support available, performance considerations, and longevity of the project. At the end of the day, choosing a framework is a nuanced process that requires careful thought and planning. At this point in time, React has quite the uptick in adoption, but as time goes by, all frameworks fall in and out of favor. What we've covered here only scratches the surface of each of these frameworks, so be sure to do your research before committing to one.

In the next chapter, we'll explore debugging JavaScript, because let's face it: we're going to make mistakes and we'll need to know how to fix them.

Further reading

- Browser wars: `https://en.wikipedia.org/wiki/Browser_wars`
- jQuery: `https://en.wikipedia.org/wiki/JQuery`
- Understanding ES6 arrow functions for jQuery developers: `https://wesbos.com/javascript-arrow-functions/`
- jQuery tutorial and reference: `https://www.w3schools.com/jquery/`
- Angular tutorial: `https://angular.io/tutorial`
- Navigating the React ecosystem: `https://www.toptal.com/react/navigating-the-react-ecosystem`
- React tutorial: `https://reactjs.org/tutorial/tutorial.html`
- Vue guide: `https://vuejs.org/v2/guide/`
- Vue comparison with other frameworks: `https://vuejs.org/v2/guide/comparison.html`

9
Deciphering Error Messages and Performance Leaks

Of course, no good language is complete without a means to detect and diagnose problems in your code. JavaScript provides rich error messages that are incredibly powerful and intuitive, but there are a few caveats and tips as you tiptoe through bug-ridden code.

As you probably know, finding a problem in your own code (a "bug") is one of the most frustrating events to occur to a developer. We pride ourselves on our code's ability to do its task, but sometimes we don't account for edge and corner cases. Additionally, error messages give us important information as we're in the process of coding by giving us important diagnostic information. Luckily, there are tools that can help us understand what's going on in our JavaScript.

Let's explore.

The following topics will be covered in this chapter:

- The Error object
- Using debuggers and other tools
- Accommodating JavaScript's performance limitations

Technical requirements

Be prepared to work through the `Chapter-9` examples from GitHub at `https://github.com/PacktPublishing/Hands-on-JavaScript-for-Python-Developers/tree/master/chapter-9/`.

We'll be working with the developer tools in the browser, and for the purposes of illustration, the instructions and screenshots will be from Google Chrome. If you're familiar with tools in another browser, however, the concepts are similar. You may also want to have a JSON parsing extension added to Chrome if you haven't already done so.

There are no specific hardware requirements for this chapter.

The Error object

Let's take a look at `https://github.com/PacktPublishing/Hands-on-JavaScript-for-Python-Developers/tree/master/chapter-9/error-object`. Open the `index.html` file and examine the JavaScript console. The first function, `typoError`, is invoked and throws a wonderful error.

It should look like this:

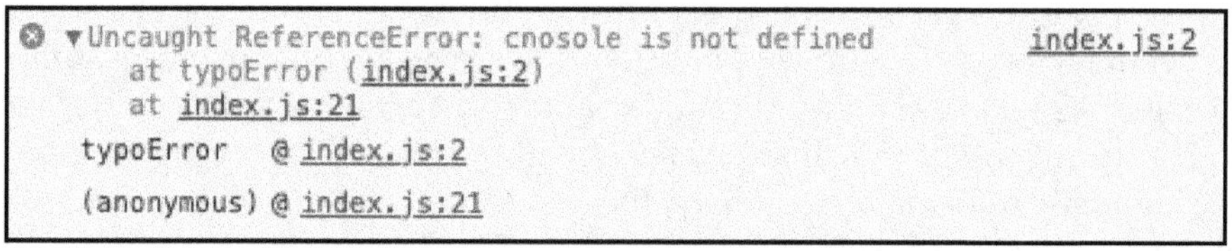

```
⊗ ▼Uncaught ReferenceError: cnosole is not defined                    index.js:2
       at typoError (index.js:2)
       at index.js:21
    typoError    @ index.js:2
    (anonymous) @ index.js:21
```

Figure 9.1 - Error console

Now, let's look at the code for our function in `index.js`:

```
const typoError = () => {
  cnosole.error('my fault')
}
```

OK! It's a simple typo, as we've all done: it should be `console.error` instead of `cnosole.error`. If you've never made a typo in code... you're a unicorn. The error message we see in the console makes it easy to see what the error is and on what line of code it lives: line 2. Now, something interesting to note is that after calling `typoError()` toward the end of the file, we also have an invocation to another function *but it doesn't fire*. We know this because (spoiler alert) it also throws errors, but we don't see them. An Uncaught ReferenceError is a **blocking error**.

In JavaScript, some errors, called blocking errors, will halt the execution of the code. Others, called **non-blocking errors**, are mitigated in such a way that the code can continue to execute even if the problem isn't resolved. There are a few ways of dealing with errors, and you should do so when faced with potential vectors for errors. Do you remember Chapter 7, *Events, Event-Driven Design, and APIs*, where we used a .catch() block in our fetch() call to gracefully handle Ajax errors? The same principle applies here. This is obviously a very contrived example, but let's go ahead and mitigate our error, like this:

```
const typoError = () => {
  try {
    cnosole.error('my fault')
  } catch(e) {
    console.error(e)
  }
}
```

Using a try/catch block for a typo is overkill, but let's pretend it was something more serious, such as an asynchronous call or a dependency from another library. If we take a look at our console output now, we'll see that our second function, fetchAttempt, has fired and it is also producing errors. Open the index-mitigated.js file and the accompanying index-mitigated.html file.

You should see this in the console of index-mitigated.html:

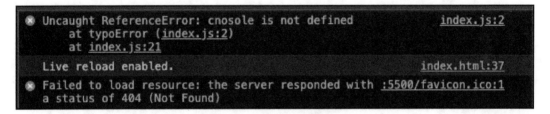

Figure 9.2 - Non-blocking error

Here, we see that our code does not stop at the typo; we've made it into a non-blocking error with our try/catch. We see that our fetchAttempt function is firing and giving us a different kind of error: 404 Not Found. Since we entered a non-existent URL (ending in undefined on purpose), we receive another error after that: a SyntaxError from our promise.

At first glance, this error may be difficult to understand, as it is explicitly talking about an unexpected character in the JSON. In Chapter 7, *Events, Event-Driven Design, and APIs*, we worked with the Star Wars API: `https://swapi.dev/`:

1. Let's look at the JSON of the example response from `https://swapi.dev/api/people/1/`. This could be a good time to ensure you have a JSON parsing extension in your browser:

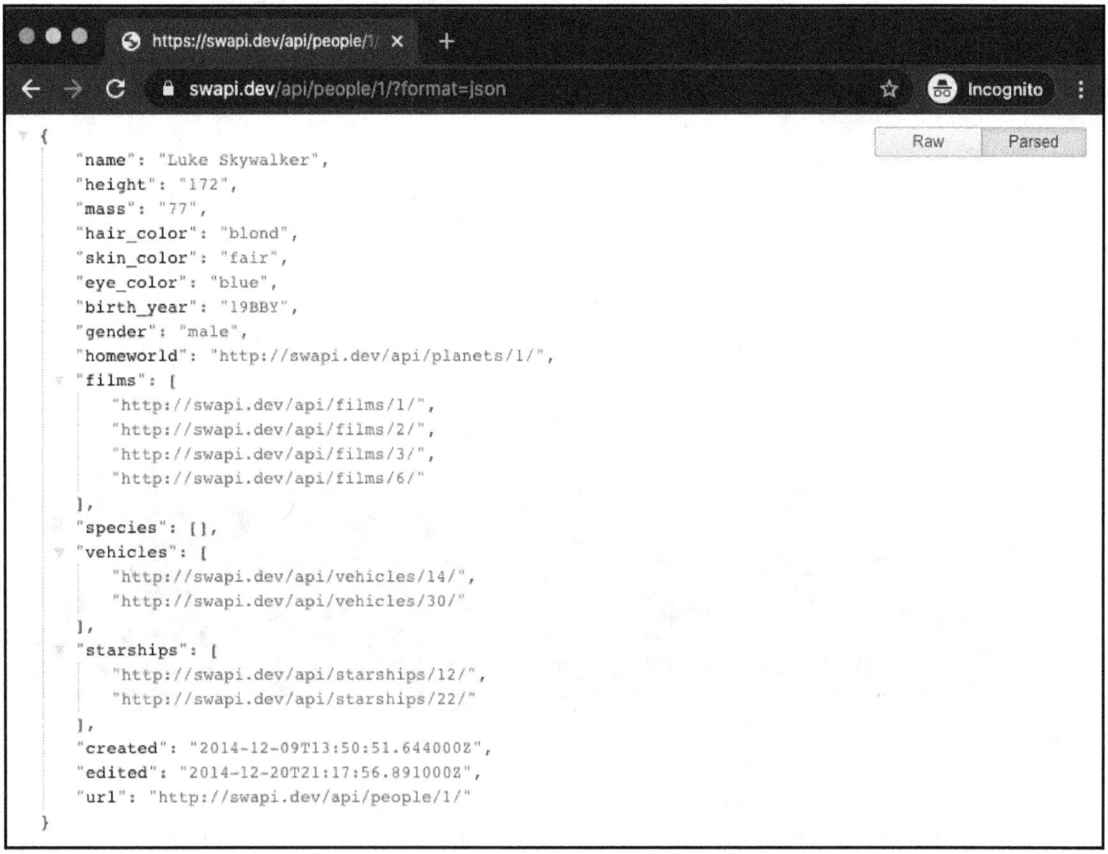

Figure 9.3 - JSON from https://swapi.dev/api/people/1/

2. It's well-formed JSON, so even though our error specifies **Syntax Error**, it's not actually the syntax of the response data that's the problem. We'll need to look deeper. Let's look at what we're getting from our `fetchAttempt` call in the Chrome JavaScript debugger. Let's click the link for the second error in our code here:

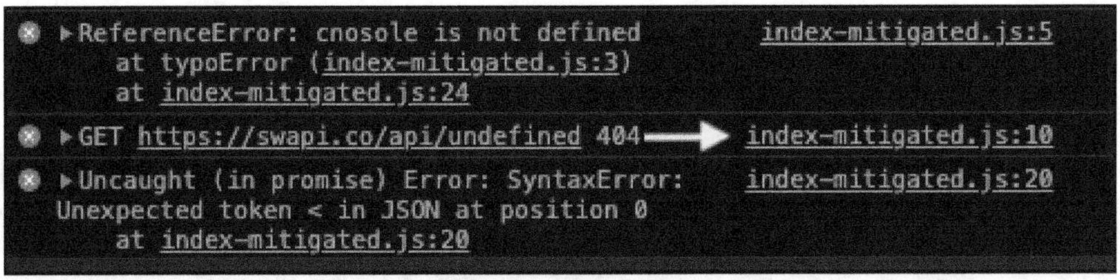

```
ⓧ ▸ReferenceError: cnosole is not defined          index-mitigated.js:5
      at typoError (index-mitigated.js:3)
      at index-mitigated.js:24
ⓧ ▸GET https://swapi.co/api/undefined 404 ━━━▶   index-mitigated.js:10
ⓧ ▸Uncaught (in promise) Error: SyntaxError:       index-mitigated.js:20
    Unexpected token < in JSON at position 0
      at index-mitigated.js:20
```

Figure 9.4 - Following the trail of the 404...

We then see this panel, with the red squiggly underlines and red markers to indicate errors:

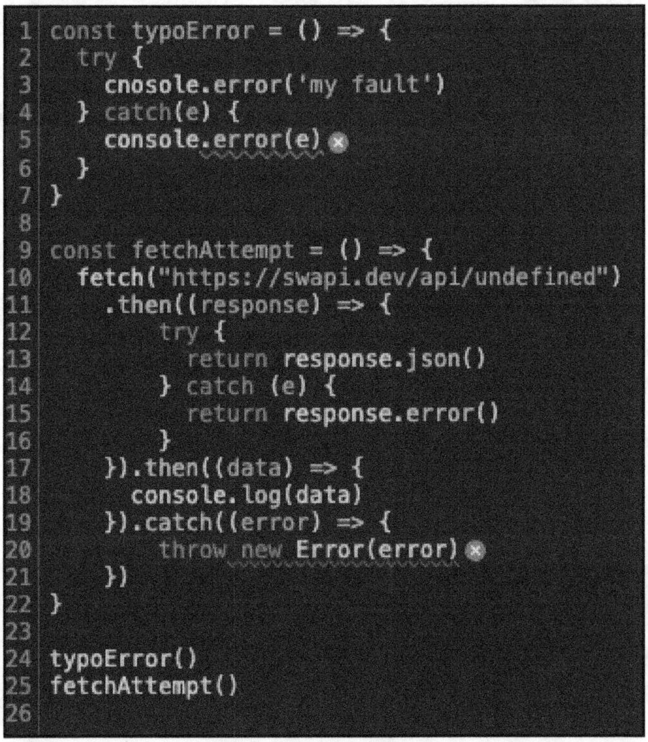

```
 1  const typoError = () => {
 2    try {
 3        cnosole.error('my fault')
 4    } catch(e) {
 5        console.error(e) ⓧ
 6    }
 7  }
 8
 9  const fetchAttempt = () => {
10    fetch("https://swapi.dev/api/undefined")
11      .then((response) => {
12        try {
13          return response.json()
14        } catch (e) {
15          return response.error()
16        }
17    }).then((data) => {
18      console.log(data)
19    }).catch((error) => {
20        throw new Error(error) ⓧ
21    })
22  }
23
24  typoError()
25  fetchAttempt()
26
```

Figure 9.5 - Errors in the debugger

3. So far, so good. If you hover over the red **X** on line 20, the tooltip advises us of the 404 error.
4. Navigate to the **Network** tab. This tool tracks incoming and outbound HTTP requests.

5. Click on the call named **undefined** and then into the **Headers** panel, like so:

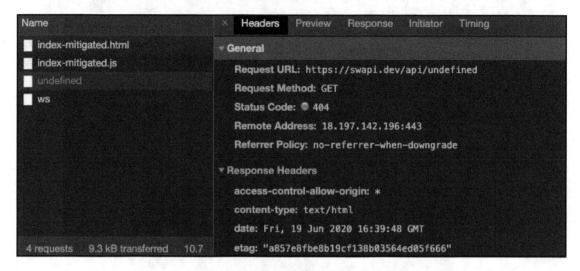

Figure 9.6 - The Headers tab

Aha! Now we see what the problem is: the JSON error is helpful, but steered us in the wrong direction. The error isn't with the JSON itself, but rather, the error means that the response *isn't JSON at all*! It is an HTML 404 error, so there is no JSON data. Our problem is confirmed to be in the URL fetching a non-existent address, and so an error page is rendered, which makes no sense to the JSON parser of `fetch`.

Let's spend some more time with debugging tools.

Using debuggers and other tools

Many web developers choose to use Google Chrome as their browser of choice as it provides a wealth of developer tools out of the box. If Chrome is not your browser of choice, here are a few browsers that have developer tools that are similar.

Safari

Safari ships with developer mode off by default, so if you use Safari, toggle the Develop menu in the **Advanced** pane in the preferences at the bottom:

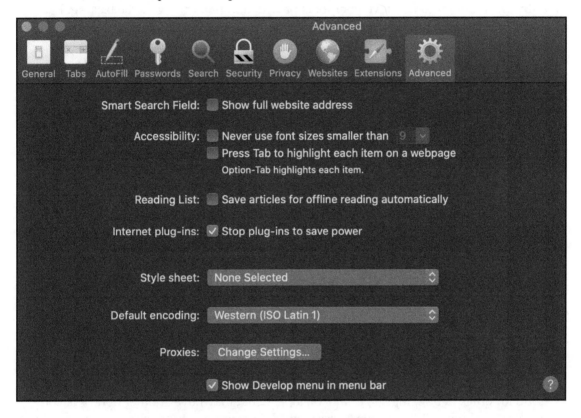

Figure 9.7 - Adding the Develop Menu to Safari

Now, you'll have a Develop menu with tools that may render error messages slightly differently than Chrome, but that are still accessible.

Internet Explorer and Microsoft Edge

With all sincerity and only a little bit of prejudice, I would recommend *not* using Internet Explorer or Microsoft Edge for JavaScript development. It is important to test your code cross-browser, but I find the developer tools provided in IE and Edge to be lacking. For example, let's take a look at the exact same page in Edge's developer tools:

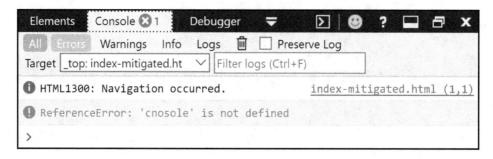

Figure 9.8 - Edge JavaScript Console

Even though we mitigated our error with a try/catch block, Edge still treats the typo as a blocking error. There are other idiosyncrasies of the Microsoft browsers as well, which date back to the browser wars we learned about before, so a good rule of thumb is to develop in Chrome and then test in a Microsoft browser to ensure cross-browser compatibility.

While there are development tools in all major browsers, the examples used here will be from Chrome. Let's take a closer look at the JavaScript console itself.

The JavaScript console

The console is not only a place to see errors but it can be used to execute code as well. This is useful for quick debugging, especially on a page that may have another library of code incorporated within it. The console has *scope* to all JavaScript loaded on the page, as long as it's accessible from the top-level `window` object. We wouldn't expect to have access to a function's internal variables, but if the browser can access the data, we can access it in the console.

Open the `fetch.html` and `fetch.js` files in the `debugger` folder and take a look. Here's the `fetch.js` file:

```
const fetchAttempt = (url) => {
  fetch(url)
    .then((response) => {
      return response
    }).then((data) => {
      if (data.status === 500) {
```

```
        console.log("We got a 500 error")
    }
    console.log(data)
    }).catch((error) => {
      throw new Error(error)
    })
}
```

It's a bare-bones `fetch` request with the URL to be fetched as a parameter to our function. In the console of our HTML page, we can actually execute this function, like so:

Figure 9.9 - Executing code in the console

As you typed in `fetchAttempt('http://httpstat.us/500')`, did you notice that the console gave you autocomplete code hints? This is another useful tool for determining whether you have access to the functions and variables at the level at which you are working. Now we see that we can execute code in the console without having to alter our JavaScript file. What did we learn from our console? Our `data.status` was indeed 500, so we threw the console error from line 7. From line 9, we got our response data, which explicitly states 500. It may go without saying but the `console.log`, `console.error`, and `console.info` functions can be invaluable as you debug JavaScript. Use them frequently, but do remember to remove them before pushing your code to a production-level environment, as they can degrade site performance if you log large objects or log too frequently.

One of the tricky things about JavaScript is that you may be dealing with hundreds of lines of code, sometimes from a third party. Luckily, the tooling of most browsers allows setting *breakpoints* in the code, which halt the execution of the code at specified points. Let's take a look at our previous file in the console and set a few breakpoints. If we click the error for line 7, the **Sources** panel is displayed. If you click a line number, you will set a breakpoint, as so:

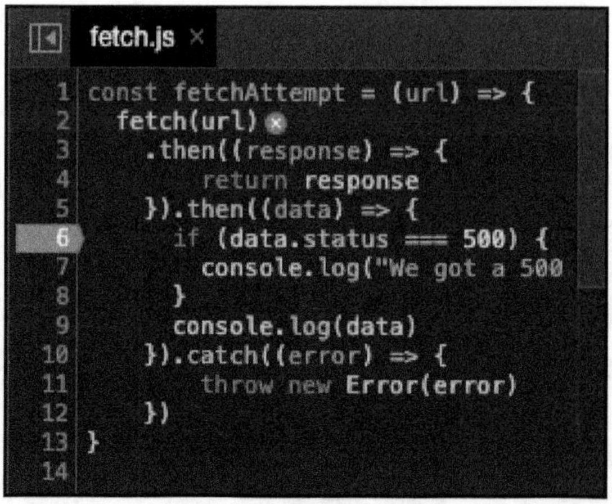

```
fetch.js  ×

1  const fetchAttempt = (url) => {
2    fetch(url) ⊗
3      .then((response) => {
4          return response
5      }).then((data) => {
6          if (data.status === 500) {
7            console.log("We got a 500
8          }
9          console.log(data)
10     }).catch((error) => {
11         throw new Error(error)
12     })
13  }
14
```

Figure 9.10 - Note the arrow marker on line 6

It's often useful to set breakpoints before the line on which the browser complained in order to more thoroughly trace through the variables being passed to our code. Let's run our code again with a page refresh and see what happens:

1. Set breakpoints on lines 6 and 7.
2. Refresh the page.
3. Navigate to the console and execute our previous command:
   `fetchAttempt('http://httpstat.us/500').`

The browser will pull up the **Sources** tab again and we should see something similar to this:

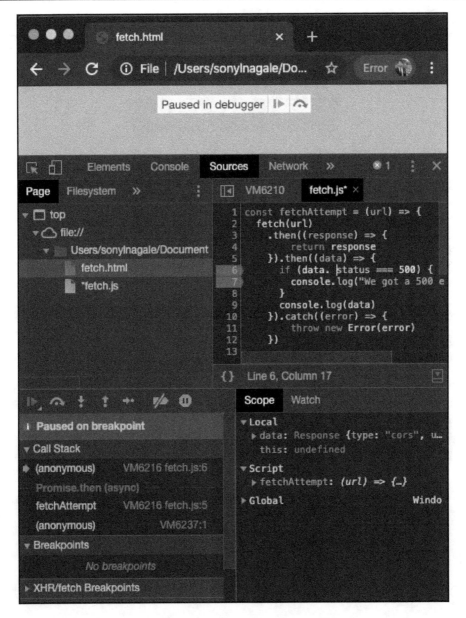

Figure 9.11 - Result of the breakpoint

We can see that in the **Scope** tab, we get a listing of the variables defined in the context within which we are executing code. If we then use the **Step** button, as shown in the screenshot, we can keep moving through our breakpoints and execute subsequent lines of code:

Figure 9.12 - The Step button

As we step through the breakpoints, the **Scope** panel will update to show our current context, which gives us more information than an explicit `console.log` function.

Let's take a look now at some ideas for how to improve your JavaScript code for performance.

Accommodating JavaScript's performance limitations

As with any language, there are ways to write JavaScript and better ways to write it. What is not as obvious in other languages, however, is the direct implications of your code for the user experience of a website. Complicated, inefficient code can clog up a browser, eat CPU cycles, and, in some cases, even crash the browser.

Take a look at this simple four-line snippet by Talon Bragg from `https://hackernoon.com/crashing-the-browser-7d540beb0478`:

```
txt = "a";
while (1) {
    txt = txt += "a"; // add as much as the browser can handle
}
```

Warning: do *not* attempt to run this in a browser! If you're curious about what this does, it will eventually create an out-of-memory exception in the browser that will kill the tab with a message that the page has become unresponsive. Why is this? Our `while` loop has a simple truthy value for its condition, so it will continue adding `"a"` to the string text until the memory allocated to that browser process is exhausted. Depending on the behavior of your browser, it may crash the tab, the whole browser, or worse. We all have experience with unstable programs (the Windows Blue Screen of Death, anyone?) but a browser failure can usually be avoided. Apart from coding best practices, such as minimizing loops and avoiding reassigning variables, there are ideas specific to JavaScript to point out. W3Schools has a few examples that are handy at `https://www.w3schools.com/js/js_performance.asp` and I'd like to underscore one specifically.

One of the most memory-intensive operations in a standard JavaScript application is DOM access. A line as simple as `document.getElementById("helloWorld")` is actually a fairly expensive operation. As a best practice, if you're going to use a DOM element more than once in your code, you should save it to a variable and act on that variable instead of going back to DOM traversal. If you think back to `Chapter 6`: *The Document Object Model (DOM)*, we stored the sticky note DOM element as a variable: `https://github.com/PacktPublishing/Hands-on-JavaScript-for-Python-Developers/blob/master/chapter-6/stickies/solution-code/script.js#L13`.

The Memory panel

Without getting into too much detail about how computers allocate memory, suffice it to say that an improperly written program can cause a memory leak by not properly releasing and recycling memory, which can cause the program to crash. As opposed to some lower-level languages, JavaScript is supposed to automatically garbage collect: the practice of automatic memory management that releases memory by destroying unneeded pieces of data. However, there are cases when improperly written code can cause a memory leak that garbage collection doesn't handle.

Since JavaScript runs client-side, it can be hard to decipher exactly what's going on in your program. Luckily, there are tools to help. Let's work through an example of a program that will allocate a lot of memory. Take a look at this example: `https://github.com/PacktPublishing/Hands-on-JavaScript-for-Python-Developers/blob/master/chapter-9/memory-leak/index.html`.

If you look at the included JavaScript file, you'll see it's very simple, yet very powerful:

```
// Based on
https://developers.google.com/web/tools/chrome-devtools/memory-problems
```

```
let x = []
const grow = (log = false) => {
  x.push(new Array(1000000).join('x'))
  if (log) {
    console.log(x)
  }
}

document.getElementById('grow').addEventListener('click', () => grow())
document.getElementById('log').addEventListener('click', () => grow(true))
```

Let's inspect our code and see what happens when we work with this simple script. Note that some of these instructions may be different depending on your browser and OS version:

1. Open the `index.html` page in Chrome.
2. Open the developer tools.
3. From the **More tools** menu, select **Performance monitor**:

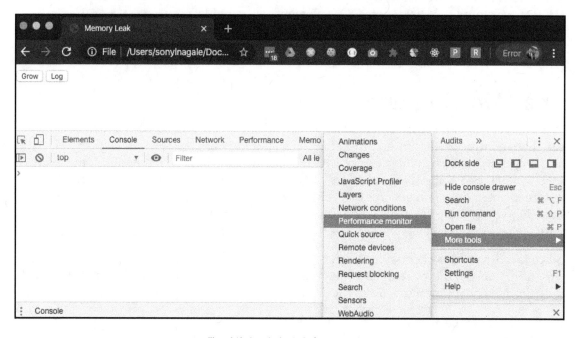

Figure 9.13 - Investigating the Performance monitor

You will see a panel with a moving timeline: `https://github.com/PacktPublishing/Hands-on-JavaScript-for-Python-Developers/blob/master/chapter-9/memory-leak/memory-leak.gif`.

4. Now, click the **Grow** button several times. You should see an increase in the JavaScript heap size, perhaps to the 13 MB range. However, as you keep clicking, the heap size shouldn't increase past where it already is.

   Why is this? In modern browsers, it's actually gotten a little more difficult to accidentally create a memory leak. In this case, Chrome is smart enough to do some trickery with memory and not cause a large increase in memory as we repeat the actions.

5. However, now start clicking the **Log** button several times. You'll see the output in the **Console** as well as an increase in the heap size:

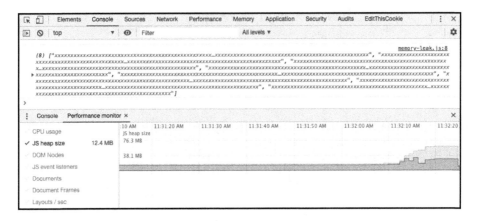

Figure 9.14 - Memory heap investigation

Notice how the graph increases in size. However, over time, the memory allocation will actually drop if you stop clicking **Log**. This is an example of Chrome's intelligent garbage collection at work.

Summary

We all make mistakes when coding, and knowing how to find, diagnose, and debug those problems is a key skill in any language. In this chapter, we've taken a look at how the Error object and the console provide us with rich diagnostic information on where an error occurred, what details are piggybacking on the object, and how to read them. Don't forget that, sometimes, the error may look one way on the surface (our JSON error in *The Error Object* section), and don't be afraid to experiment with tracing through your code with console statements and breakpoints.

Since JavaScript runs client-side, it's important to keep in mind the performance capacity of your users. There are many best practices when writing JavaScript, such as reusing variables (especially DOM-related ones), so always be sure to keep your code **DRY** (**Don't Repeat Yourself**).

In the next chapter, we'll wrap up working with the frontend and understand how JavaScript truly is the ruler of the frontend.

Questions

1. What is the root cause of memory problems?
    1. The variables in your program are global.
    2. Inefficient code.
    3. JavaScript's performance limitations.
    4. Hardware inadequacies.

2. When using DOM elements, you should store references to them locally versus always accessing the DOM.
    1. True
    2. False
    3. True when using them more than once

3. JavaScript is pre-processed on the server side, and thus more efficient than Python.
    1. True
    2. False

4. Setting breakpoints can't find memory leaks.
    1. True
    2. False

5. It's a good idea to store all variables in the global namespace as they're more efficient to reference.
    1. True
    2. False

Further reading

For more information, you can use the following links:

- Isolating Memory Leaks with Chrome's Allocation Timeline: `https://blog.logrocket.com/isolating-memory-leaks-with-chromes-allocation-timeline-244fa9c48e8e/`
- Garbage Collection: `https://en.wikipedia.org/wiki/Garbage_collection_(computer_science)`
- JavaScript Performance: `https://www.w3schools.com/js/js_performance.asp`
- Memory Problems: `https://developers.google.com/web/tools/chrome-devtools/memory-problems`
- Node.js Memory Leak Detection: `https://medium.com/tech-tajawal/memory-leaks-in-nodejs-quick-overview-988c23b24dba`

10
JavaScript, Ruler of the Frontend

If you're starting to grasp that JavaScript is integral to the way modern websites and web applications function, you're on the right path. Without JavaScript, most of the user interfaces we take for granted on the web wouldn't exist. Let's take a closer look at how JavaScript brings the frontend together. We'll be working with a couple of React applications, as well as comparing and contrasting a Python application in order to further our understanding of the whys and hows of JavaScript's importance on the frontend.

The following topics will be covered in this chapter:

- Building interactions
- Using dynamic data
- Understanding modern applications

Technical requirements

Be prepared to work with the code provided in the `Chapter-10` directory of the repository at `https://github.com/PacktPublishing/Hands-on-JavaScript-for-Python-Developers/tree/master/chapter-10`. As we'll be working with command-line tools, also have your Terminal or command-line shell available. We'll need a modern browser and a local code editor.

Building interactions

Let's take a look at a simple **Single-Page Application** (**SPA**):

1. Navigate to the `simple-reactjs-app` directory in `chapter-10` (`cd simple-reactjs-app`).
2. Install the dependencies with `npm install`.
3. Run the app with `npm start`.
4. Access the app at `http://localhost:3000`. Here's what you'll see:

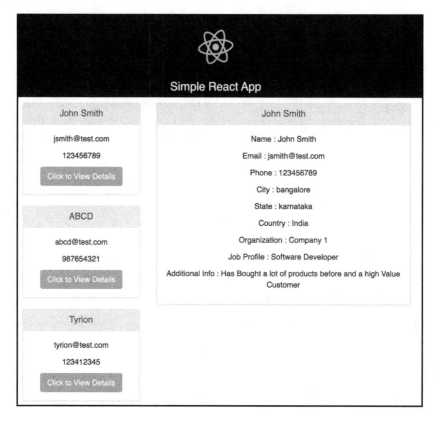

Figure 10.1 – Simple React app

As you click through the detail buttons and inspect your network tab, you will see that the page does not reload and it only loads the JSON data from the server. This is a very basic example of how an SPA functions: with minimal server usage, the user experience's interactions are streamlined, contributing to an efficient, low-overhead workflow. You're probably familiar with other SPAs, such as Gmail, Google Maps, and Facebook, though the underlying technologies vary.

It may be taken for granted in this day and age of internet technology, but JavaScript is the foundation of how these apps work. Without JavaScript, we'd have a lot of page reloading and high wait-times, even with Ajax.

Let's see how we can work with dynamic data by comparing and contrasting a basic Python example with a modern React application.

Using dynamic data

Let's first take a look at a Python Flask example:

1. Navigate to the `flask` directory in `chapter-10` (`cd flask`).
2. You'll need to install a few pieces for our setup. These instructions are for Python:
    1. Create a virtual environment with `python3 -m venv env`.
    2. Activate it using `. env/bin/activate`.
    3. Install the requirements: `pip3 install -r requirements.txt`.
    4. Now you can start the application: `python3 app.py`.
3. Access the page at `http://localhost:5000`. You'll see this:

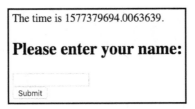

Figure 10.2 – Basic Flask application

Play around with entering and not entering your name and take a look at the fact that the page reloads when you do so (I've added a timestamp to make it easier to see as the page reload can happen too quickly to see). This is a very basic Flask application, and there are more efficient ways to do validation work with a combination of Python and JavaScript, but at a base level, even with some Flask-based form validation tools, the validation and interaction we're seeing happens on the backend. The server is hit every time we hit **Submit**. The following screenshot shows server-side validation if you don't enter a string:

Figure 10.3 - Basic Flask validation

Note that the timestamp changes, indicating a re-render from the server.

Let's make this a bit better for our form validation interaction by revising our simple React application:

1. Navigate to the `reactjs-app-form` directory: `cd reactjs-app-form`.
2. Install the dependencies: `npm install`.
3. Start the server: `npm start`.
4. Access the page at `http://localhost:5000`. Here's an updated version of our simple app:

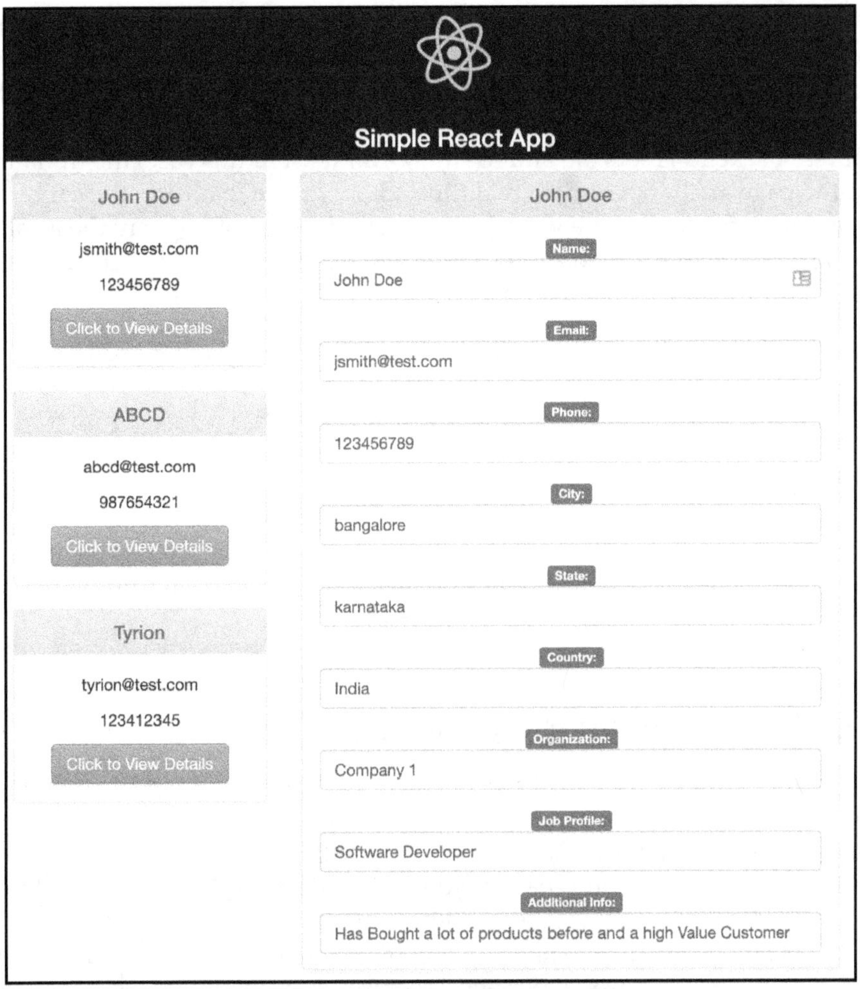

Figure 10.4 – Simple app with dynamic data

Now try working with it and notice how if you change one of the primary fields, the fields on the left will change too. Additionally, it's saving the JSON *as you edit* so that if you refresh the page, your changes persist. This is thanks to the power of JavaScript: the React frontend is handing all the changes you're making throughout parts of the application and then an Express backend is serving and saving JSON files. In this case, the updates to the markup on the page are happening in real time. Granted, we're hitting the server with a save and read operation each time we edit, but that's because of the way the application is designed. To persist changes, it would be trivial to create a **Save** button instead of the saving happening on the change of a field.

If you'd like to work with this example, there are a few things you'll need to do:

1. First, navigate to the directory in a new shell window (leave the previous instance running): `cd client`.
2. Execute `npm install`.
3. Begin the program: `npm start`.

The Express server will then gather the built files created by React's running process versus in the pre-built files that are already in the directory.

Input validation and error handling

One part about dynamic data that we can see that might be important for an application such as this is *input validation* and *error handling*. Notice how in our application, if the email field is blank or we haven't entered a valid email, it will have a red outline. Otherwise, it will have a green outline. As you type a valid email address and select the next field, you can see that the red outline will change to green without hitting the server (except to save the data, as we discussed before). This is client-side validation, and it's extremely powerful when creating a streamlined user experience: the user does not have to hit save and wait for a server response in order to see whether they've entered incorrect data.

You might have noticed one detail when working with the **Phone** field: it's restricted to numbers. If you look at `client/src/CustomerDetails.js`, we've restricted the type to numbers here:

```
<Input name="phone" type="number"
value={this.state.customerDetails.data.phone || ''}
onChange={this.handleChange} />
```

There are a couple of other React pieces in here also. Let's take a look at the `handleChange` function:

```
handleChange(event) {
    const details = this.state.customerDetails
    details.data[event.target.name] = event.target.value
    this.validate(event.target)

    this.setState({ customerDetails: details })
    console.log(this.state.customerDetails)

    axios.post(`${CONSTANTS.API_ROOT}/api/save/` +
    this.state.customerDetails.data.id, details)
      .then(() => {
        this.props.handler();
      })
}
```

Axios is a library that simplifies Ajax calls, and I used it here instead of `fetch` just as a demonstration. You will probably see Axios being used in React work, although you can always choose to use raw `fetch`. However, let's focus on the `this.validate(event.target)` line.

This is the content of the function:

```
validate(el) {
    const properties = (el.name) ? el : el.props

    if (properties.name === 'email') {
      if (validateEmail(properties.value)) {
        this.setState({ validate: { email: true }});
      } else {
        this.setState({ validate: { email: false }});
      }
    }
}
```

`validateEmail()` is a magic function! You can find it in `client/src/validation.js` and it uses a *regular expression* to pattern-match the input string to see whether it looks like a correctly formatted email address. Then, depending on whether the function returns `true` or `false`, we set a validation state that React will then use to set the color of the email field's border.

Frontend validation and error handling are incredibly important for a smooth user experience, but it's only part of the story. The other part is security.

Security and data

As you know from working with the developer tools in the browser, it's possible to circumvent pretty much any frontend restrictions if you try hard enough. For example, with our **Phone** field, though we've restricted it on the frontend, we can always inspect the HTML and enter in any value that we want. As a quick note, it's important to also validate your data on the backend to be sure it's properly formatted.

One thing that corporate data breaches and hacks always have in common is that the attacker is taking advantage of a weakness in the system being used. Very rarely is it actually a case of a leaked password; more often, it's weak cryptography or even a frontend problem. We'll discuss security further in Chapter 17, *Security and Keys*. You can learn more at OWASP.org.

Let's continue and review what we've learned.

Understanding modern applications

At this point, it should come as no surprise that all modern web applications are inextricably tied to using JavaScript. There's no modern web without it, as interactions simply cannot happen in real time without it. The server side has its place and importance, but the crux of what the user sees and interacts with is controlled by JavaScript.

Just like CSS is a complement to HTML, JavaScript is the third friend in the group, creating meaningful experiences out of a collection of markup and styling. As the muscles of a web app, it provides us with rich interactions and logic, and is the foundation of all SPAs. It truly is a tool of wonder and beauty.

Summary

With JavaScript, we can go beyond the "web page" and create fully fledged web applications. From email systems to banks to spreadsheets to nearly anything you use a computer for, JavaScript is there to help you.

In the next chapter, we will venture into using JavaScript on the server side with Node.js. We won't completely leave the frontend behind, but rather, we'll see how they tie together.

3
Section 3 - The Back-End: Node.js vs. Python

Now that we've seen how JavaScript can be used on the front-end (a new and possibly scary place), let's switch gears and go to the back-end. Node.js has more in common with Python when used on the back-end, but there are significant differences. Let's build an application to explore how Node.js works.

In this section, we will cover the following chapters:

- Chapter 11, *What Is Node.js?*
- Chapter 12, *Node.js vs. Python*
- Chapter 13, *Using Express*
- Chapter 14, *React with Django*
- Chapter 15, *Combining Node.js with the Frontend*
- Chapter 16, *Enter Webpack*

11
What Is Node.js?

Now that we've examined the use of JavaScript on the frontend, let's dive into its role in the "JavaScript everywhere" paradigm using Node.js. We discussed Node.js in Chapter 2, *Can We Use JavaScript Server-Side? Sure!*, so now it's time to dive deeper into how we can use it to create rich server-side applications.

The following topics will be covered in this chapter:

- History and usage
- Installation and usage
- Grammar and structure
- Hello, World!

Technical requirements

Be prepared to work with the code provided in the Chapter-11 directory of the repository: https://github.com/PacktPublishing/Hands-on-JavaScript-for-Python-Developers/tree/master/chapter-11. As we'll be working with command-line tools, also have your Terminal or command-line shell available. We'll need a modern browser and a local code editor.

History and usage

First released in 2009, Node.js has been widely adopted in the industry by major corporations and smaller companies alike. There are literally thousands of packages available for use in Node.js, creating a rich ecosystem of users and a community of developers. As with any open source project, community support is crucial to the adoption and longevity of the technology.

From a technical standpoint, Node.js is a runtime environment in a single-threaded event loop. What this means in practice is that it can handle thousands upon thousands of concurrent connections without the overhead of switching between contexts. For those who are more familiar with other architectural patterns, a single thread might seem counterintuitive, and it used to be held up as an example of Node.js's perceived breakpoint. However, it can be argued that the stability and reliability of a Node.js system has shown this paradigm to be sustainable. There are ways to augment a server's capacity for handling requests, but it should be noted that it's a bit more nuanced than just throwing additional hardware resources at the problem. How to scale Node.js is a bit out of scope for this book, but there are techniques involving the underlying library, called libuv.

At the time of writing, the biggest feather in Node.js's cap might be its powering of Twitter. According to SimilarTech, its 4.3 billion monthly visits stand as a testament to its power. Now, I'm sure the Twitter team has done some incredible architecting over the years to power the platform, and we rarely see the famous Twitter "fail whale" anymore; I would argue that the reliance on Node.js is a good thing that has helped provide sustainability and reliability.

Onward to using it!

Installation and usage

The easiest way to install Node.js is to use the installers provided for you at `https://nodejs.org`. These packages will guide you through the installation of Node.js on your system. Be sure to also install `npm`, Node's package manager. You can refer to `Chapter 3`, *Nitty-Gritty Grammar*, for more details on installation.

Let's give it a go:

1. Open a Terminal window.
2. Type `node`. You will see a simple > to indicate that Node.js is running.
3. Type `console.log("Hi!")` and hit *Enter*.

It's really as simple as that! Exit the command prompt either by hitting *Ctrl* + *C* twice or typing `.exit`.

So, that's fairly basic. Let's do something a bit more interesting. Here's the contents of `chapter-11/guessing-game/guessing-game.js`:

```
const readline = require('readline')
const randomNumber = Math.ceil(Math.random() * 10)
```

```
const rl = readline.createInterface({
  input: process.stdin,
  output: process.stdout
});

askQuestion()

function askQuestion() {
  rl.question('Enter a number from 1 to 10:\n', (answer) => {
    evaluateAnswer(answer)
  })
}

function evaluateAnswer(guess) {
  if (parseInt(guess) === randomNumber) {
    console.log("Correct!\n")
    rl.close()
    process.exit(1)
  } else {
    console.log("Incorrect!")
    askQuestion()
  }
}
```

Run the program with `node guessing-game.js`. As you can probably tell from the code, the program will select a random number between 1 and 10 and then ask you to guess it. You can enter numbers at the command prompt in order to guess the number.

Let's break down this example in the next section.

Grammar and structure

The great thing about Node.js is that you already know how to write it! Take this for example:

JavaScript	Node.js
`console.log("Hello!")`	`console.log("Hello!")`

That's not a trick: it's identical. Node.js is syntactically almost identical to browser-based JavaScript, right down to the fight between ES5 and ES6, as we've discussed previously. In my experience, there is still a preponderance of ES5-style code in use with Node.js, so you will see code with `var` instead of `let` or `const`, as well as a healthy use of semicolons. You can review Chapter 3, *Nitty-Gritty Grammar* for more information on these distinctions.

In our guessing game example, we see one thing that is new to us – the first line:

```
const readline = require('readline')
```

Node.js is a *modular* system, which means that not all parts of the language will be brought in at once. Rather, modules will be included when the `require()` statement is issued. Some of these modules will be built-in to Node.js, as `readline` is, and some will be installed via npm (more to come on that part). We use the `readline.createInterface()` method to create a way to use our input and output, and then the rest of the code of our guessing game program should make some sense. It's simply going to ask the question over and over until the number entered equals the random number generated by the program:

```
function evaluateAnswer(guess) {
  if (parseInt(guess) === randomNumber) {
    console.log("Correct!\n")
    rl.close()
    process.exit(1)
  } else {
    console.log("Incorrect!")
    askQuestion()
  }
}
```

Let's take a look at an example that reads files from the filesystem, which is something we can't do from an ordinary client-side web application.

Customer lookup

Take a look at the customer-lookup directory, `https://github.com/PacktPublishing/Hands-on-JavaScript-for-Python-Developers/tree/master/chapter-11/customer-lookup`, and run the script with `node index.js`. It's fairly simple:

```
const fs = require('fs')
const readline = require('readline')

const rl = readline.createInterface({
  input: process.stdin,
  output: process.stdout
});

const customers = []

getCustomers()
ask()
```

```
function getCustomers() {
 const files = fs.readdirSync('data')

 for (let i = 0; i < files.length; i++) {
   const data = fs.readFileSync(`data/${files[i]}`)
   customers.push(JSON.parse(data))
 }
}

function ask() {
 rl.question(`There are ${customers.length} customers. Enter a number to
 see details:\n`, (customer) => {
   if (customer > customers.length || customer < 1) {
     console.log("Customer not found. Please try again")
   } else {
     console.log(customers[customer - 1])
   }
   ask()
 })
}
```

Some of this will look familiar, like the `readline` interface. Here's something new that we're working with, though: `const fs = require('fs')`. This is bringing in the filesystem module so that we can work with files stored on the filesystem. If you look in the data directory, you will find four basic JSON files.

We're doing three things in the `getCustomers()` function:

1. Using `readdirSync` to get a list of files in the data directory. When working with filesystems, you can interact with the system in a synchronous or asynchronous manner, similar to interacting with APIs and Ajax. For ease of use in this example, we'll be working with the synchronous filesystem calls.
2. Now `files` will be a listing of the files in the data directory. Loop through the files and store the contents in the `data` variable.
3. Push the parsed JSON into the `customers` array.

So far so good. The `ask()` function should also be easy to understand, as we're just seeing whether the number entered by the user exists in the array and then returning the data in the associated file.

Now let's look at how we can use an open source project in Node.js to achieve a (rather silly) goal: creating a text-art representation of a photo.

ASCII art and packages

We'll be working with the instructions in the GitHub repository at `https://www.npmjs.com/package/asciify-image`:

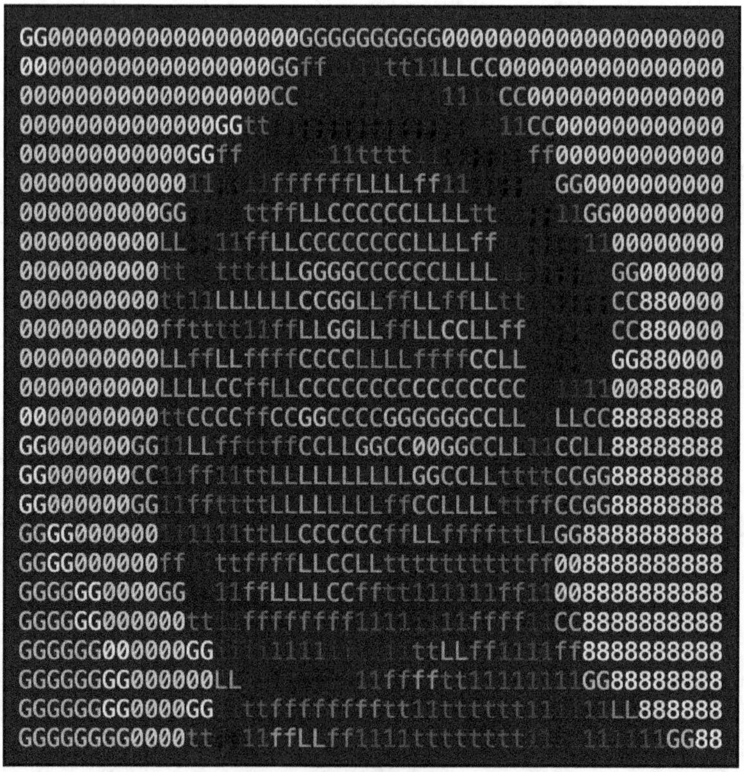

Figure 11.1 - An ASCII art representation of me!

Here's the installation step by step:

1. Create a new directory called `ascii-art`.
2. `cd ascii-art`
3. `npm init`. You can accept the defaults provided by npm.
4. `npm install asciify-image`

Now, let's have some fun:

1. Place an image in the `ascii-art` directory, such as a JPEG sized to no more than 200 x 200 pixels or so. Name it `image.jpg`.
2. Create `index.js` in the directory and open it.
3. Enter this code:

```
const asciify = require('asciify-image')

asciify(__dirname + '/image.jpg', { fit: 'box', width: 25, height:
25}, (err, converted) => {
  console.log(err || converted)
})
```

4. Execute the program with `node index.js` and view your wonderful artwork! Depending on your terminal colors, you may have to work with some of the options to change colors around to display on a light background. These are documented in the GitHub repository linked previously.

What have we shown here? First, we used npm to initialize a project and then install a dependency. If you noticed, running these created some files and directories for you. Your directory structure should look close to this:

```
.
├── image.jpg
├── index.js
├── node_modules

├── package-lock.json
└── package.json
```

The `node_modules` directory will have a lot more files inside it. If you're familiar with source control such as Git, you'll know that the `node_modules` directory should always be *ignored* and not committed to source control.

Let's take a look at `package.json`, which will look similar to this:

```
{
  "name": "ascii-art",
  "version": "1.0.0",
  "description": "",
  "main": "index.js",
  "dependencies": {
    "asciify-image": "^0.1.5"
  },
```

```
"devDependencies": {},
"scripts": {
  "test": "echo \"Error: no test specified\" && exit 1"
},
"author": "",
"license": "ISC"
}
```

If we dissect this a bit, we'll find that this npm entry point into our program is actually rather simple. There's some metadata about the project, an object of dependencies with their version, and some scripts that we can use to control our project.

If you're familiar with npm, you may have used the `npm start` command to run a project instead of manually entering `node`. However, in our `package.json`, we don't have a start script. Let's add one.

Modify the `scripts` object to look like this:

```
"scripts": {
  "test": "echo \"Error: no test specified\" && exit 1",
  "start": "node index.js"
},
```

Don't forget to pay attention to your commas, as this is valid JSON and will break if commas are improperly used. Now, to start our program, we only have to type `npm start`.

This is a very basic example of npm scripts. It is conventional in Node.js to use `package.json` to control all of the scripts for building and testing. You can name your commands as you'd like and execute them like this: `npm run my-fun-command`.

For our next trick, we'll create a "Hello, World!" application from scratch. It will, however, do a bit more than just say hello.

Hello, World!

Create a new directory called `hello-world` and initialize a node project with `npm init`, similar to how we did previously. In Chapter 13, *Using Express*, we'll work with Express, a popular web server for Node.js. However, for now, we'll use a very bare-bones method of creating a page.

Start off your `index.js` script as follows:

```
const http = require('http')

http.createServer((req, res) => {
  res.writeHead(200, {'Content-Type': 'text/plain'})
  res.end("Hello, World!")
}).listen(8080)
```

As with `fs` and `readline`, `http` is built in to Node, so we don't have to use `npm install` to get it. Rather, this will work out of the box. Add a start script in your `package.json` file:

```
"scripts": {
    "test": "echo \"Error: no test specified\" && exit 1",
    "start": "node index.js"
},
```

Then fire it up!

```
→  hello-world git:(master) ✗ npm start

> hello-world@1.0.0 start /Users/sonylnagale/Documents/Packt/code-snippets/Hands-on-JavaScript-for-Python-Developers/chapter-11/hello-world
> node index.js
```

Figure 11.2 - Executing npm start

OK, our output isn't super helpful, but if we read our code, we can see that we've done this: "Create an HTTP server listening on port 8080. Send a 200 OK message and output 'Hello, World!'". Let's now pull up a browser and go to `http://localhost:8080`. We should see a simple page greeting us.

Great! Easy enough so far. Stop your server with *Ctrl + C* and let's code some more.

What if we could use the ASCII art generator that we used in the previous example to ask the user for input and then display the image in the browser? Let's try it.

First, we need to run `npm install asciify-image`, and then let's try this code:

```
const http = require('http')
const asciify = require('asciify-image')

http.createServer((req, res) => {
  res.writeHead(200, {'Content-Type': 'text/html'})
  asciify(__dirname + '/img/image.jpg', { fit: 'box', width: 25, height: 25
    }, (err, converted) => {
    res.end(err || converted)
  })
}).listen(8080)
```

It's similar to what we did previously to output to the command line, but we're using the `http` server `res` object to send a reply. Start your server with `npm start` and let's see what we get:

```
[38;5;188mG [39m [38;5;188mG [39m [38;5;188mG [39m [38;5;188mG [39m [38;5;188mG [39m [38;5;188mG [39m [38;5;188mG
[38;5;188mG [39m [38;5;188mG [39m [38;5;188mG [39m [38;5;188mG [39m [38;5;188mG [39m [38;5;188mG [39m [38;5;188mG
[38;5;188mG [39m [38;5;188mG [39m [38;5;188mG [39m [38;5;188mG [39m [38;5;188mG [39m [38;5;188mG [39m [38;5;188mG
[38;5;188mG [39m [38;5;188mG [39m [38;5;188mG [39m [38;5;188mG [39m [38;5;188mG [39m [38;5;188mG [39m [38;5;188mG
[39m [38;5;16m
[39m [38;5;16m. [39m [38;5;16m. [39m [38;5;59m, [39m [38;5;16m. [39m [38;5;16m. [39m [38;5;59m. [39m [38;5;59m, [39m [38;5
[38;5;188mG [39m [38;5;188mG [39m [38;5;188mG [39m [38;5;188mG [39m [38;5;188mG [39m [38;5;188mG
[39m [38;5;16m [39m [38;5;16m [39m [38;5;16m
[39m [38;5;16m. [39m [38;5;16m. [39m [38;5;52m. [39m [38;5;59m, [39m [38;5;59m, [39m [38;5;59m, [39m [38;5;59m, [39m [38;5
[39m [38;5;16m. [39m [38;5;16m. [39m [38;5;16m
[39m [38;5;16m. [39m [38;5;59m, [39m [38;5;59m, [39m [38;5;59m, [39m [38;5;59m, [39m [38;5;102m1 [39m [38;5;188mC [39m
[38;5;188mG [39m [38;5;188mG [39m [38;5;188mG [39m [38;5;188mG [39m [38;5;188mG [39m [38;5;188mG [39m [38;5;188mG
[39m [38;5;16m
[39m [38;5;16m. [39m [38;5;16m. [39m [38;5;59m, [39m [38;5;59m: [39m [38;5;144mt [39m [38;5;224m0 [39m [38;5;188m0 [39m
[38;5;188mG [39m [38;5;188mG [39m [38;5;188mG [39m [38;5;188mG [39m [38;5;188mG [39m [38;5;188mG [39m [38;5;188mG
[39m [38;5;16m. [39m [38;5;59m: [39m [38;5;101mi [39m [38;5;137mi [39m [38;5;137m1 [39m [38;5;138mt [39m [38;5;138mt [39m
[39m [38;5;16m. [39m [38;5;59m, [39m [38;5;102mt [39m [38;5;188mG [39m [38;5;224m0 [39m [38;5;188m0 [39m [38;5;188m0 [39
[38;5;188mG [39m [38;5;188mG [39m [38;5;188mG [39m [38;5;188mG [39m [38;5;188mG [39m [38;5;188mG [39m [38;5;188mG [3
[39m [38;5;16m. [39m [38;5;16m. [39m [38;5;16m. [39m [38;5;95m; [39m [38;5;137mi [39m [38;5;137m1 [39m [38;5;137m1 [39m [3
[39m [38;5;16m. [39m [38;5;59m; [39m [38;5;145mf [39m [38;5;188mG [39m [38;5;230m0 [39m [38;5;188m0 [39m [38;5;188m0 [39
[38;5;188mG [39m [38;5;188mG [39m [38;5;188mG [39m [38;5;188mG [39m [38;5;188mG [39m [38;5;188mG [39m [38;5;188mG [3
[38;5;188mG [39m [38;5;188mG [39m [38;5;188mG [39m [38;5;188mG [39m [38;5;188mG [39m [38;5;188mG [39m [38;5;188mG [3
[39m [38;5;16m. [39m [38;5;52m. [39m [38;5;95m; [39m [38;5;101m; [39m [38;5;95m; [39m [38;5;101mi [39m [38;5;138m1 [39m [3
[39m [38;5;16m. [39m [38;5;59m, [39m [38;5;181mC [39m [38;5;230m8 [39m [38;5;188m0 [39m [38;5;188m0 [39m [38;5;188m0 [39
[38;5;188mG [39m [38;5;188mG [39m [38;5;188mG [39m [38;5;188mG [39m [38;5;188mG [39m [38;5;188mG [39m [38;5;188mG [3
```

Figure 11.3 - Raw output

OK, that's not even close to what we wanted to see. Here's the rub: what we sent to the browser was *ANSI-encoded text,* not actual HTML. We'll have to do a little work to convert it. Quit the server again and...

One moment. Why do we have to keep starting and stopping the server? It turns out we *don't really* have to. There are tools to reload our server when the file changes. Let's install one called **supervisor**:

1. `npm install supervisor`
2. Modify your `package.json` **start script to read** `supervisor index.js`.

Now start your server with `npm start` and, as you code, the server will restart itself upon saving, making development much faster.

Back to the code. What we're going to need is a package to convert ANSI to HTML. Install `ansi-to-html` with `npm install` and let's get going:

```
const http = require('http')
const asciify = require('asciify-image')
const Convert = require('ansi-to-html')
const convert = new Convert()

http.createServer((req, res) => {
  res.writeHead(200, {'Content-Type': 'text/html'})
  asciify(__dirname + '/img/image.jpg', { fit: 'box', width: 25, height: 25
    }, (err, converted) => {
    res.end(convert.toHtml(err || converted))
  })
}).listen(8080)
```

If you refresh the browser, you'll see that we're getting closer!

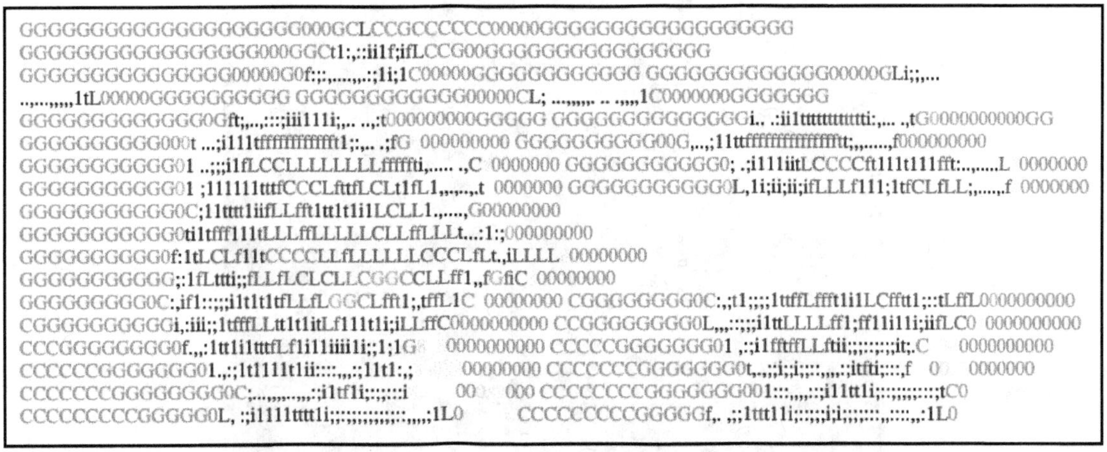

Figure 11.4 - It's HTML!

Now we really just need a little CSS:

```
const css = `
<style>
body {
  background-color: #000;
}
* {
```

```
  font-family: "Courier New";
  white-space: pre-wrap;
}
</style>
`
```

Add that in our `index.js` and concatenate it to the output, as follows:

```
asciify(__dirname + '/img/image.jpg', { fit: 'box', width: 25, height: 25
}, (err, converted) => {
  res.write(css)
  res.end(convert.toHtml(err || converted))
})
```

Now refresh and we should see our image!

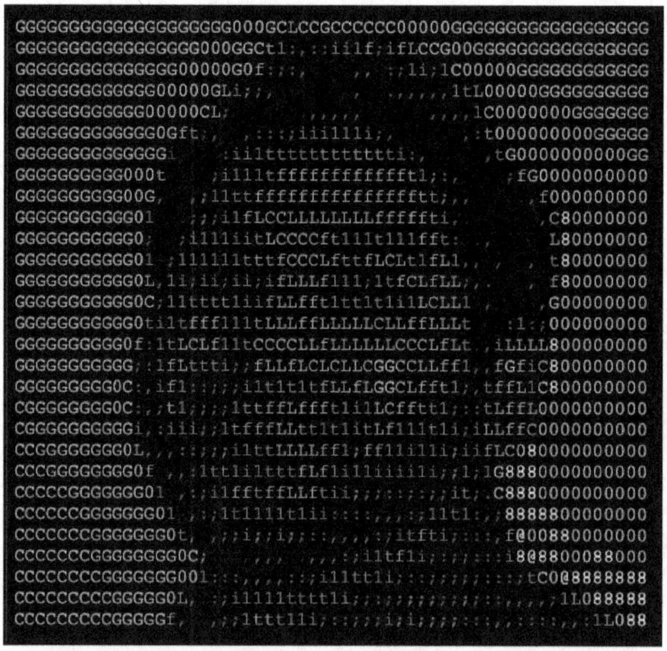

Figure 11.5 - ANSI to HTML

Fantastic! It's a little more exciting than just printing out "Hello, World!", don't you think?

Let's build on our Node.js skills by revisiting our Pokémon game from Chapter 7, *Events, Event-Driven Design, and APIs*, but this time, in Node.js.

Pokéapi, revisited

We're going to make a little terminal **command-line interface** (**CLI**) game using the Pokéapi (`https://pokeapi.co`). Since we have the basic logic of the game at `https://github.com/PacktPublishing/Hands-on-JavaScript-for-Python-Developers/tree/master/chapter-7/pokeapi/solution-code`, we're just going to get the beginnings going and then you can finish the game as a challenge by porting over the logic from the frontend to the backend with Node.js.

Start with a fresh directory and begin a new project, as follows:

1. `mkdir pokecli`
2. `npm init`
3. `npm install asciify-image axios terminal-kit`
4. Copy the Pokéapi logo from `https://pokeapi.co` to a new `img` directory with **Save Image** in the browser.
5. Make a new `index.js` file.
6. Modify `package.json` with a start script, as we've done previously: `"start"`: `"node index.js"`.

Your file structure should look like this, minus the `node_modules` directory:

```
.
├── img
│   └── pokeapi_256.png
├── index.js
├── package-lock.json
└── package.json
```

Let's begin working on our `index.js`. First off, we need to include the packages we're using:

```
const axios = require('axios')
const asciify = require('asciify-image')
const term = require('terminal-kit').terminal
```

Next, since we're going to use the API to retrieve and store our Pokémon, let's create a new object to store them at the top level so that we'll have access to them:

```
const pokes = {}
```

Now we're going to work with Terminal Kit (`https://www.npmjs.com/package/terminal-kit`) to create a better CLI experience than the standard `console.log` output and `readline` input:

```
function terminate() {
  term.grabInput(false);
  setTimeout(function () { process.exit() }, 100);
}
term.on('key', (name, matches, data) => {
  if (name === 'CTRL_C') {
    terminate();
  }
})
term.grabInput({ mouse: 'button' });
```

What we're doing here is first creating a terminate function that will exit our Node.js program after stopping `term` from capturing input, for cleanup purposes. The next method specifies that when we hit *Ctrl + C*, the program will call the `terminate()` function to exit. *This is an important part of our program, as* `term` *does not exit with Ctrl + C by default.* Lastly, we tell `term` to capture input.

To start our game, begin with a splash screen of the Pokéapi logo:

```
term.drawImage(__dirname + '/img/pokeapi_256.png', {
  shrink: {
    width: term.width,
    height: term.height * 2
  }
})
```

We can do this directly using `term` instead of the `asciify-image` library (don't worry, we'll use that later):

Figure 11.6 - Pokéapi splash screen

Next, write a function to retrieve information from the API using the Axios Ajax library:

```
async function getPokemon() {
  const pokes = await axios({
    url: 'https://pokeapi.co/api/v2/pokemon?limit=50'
  })

  return pokes.data.results
}
```

Axios (`https://www.npmjs.com/package/axios`) is a package to make requests easier than `fetch` by reducing the number of promises required. As we saw in previous chapters, `fetch` is powerful, but does require a bit of chaining of promise resolutions to operate. This time, let's use Axios. Note that the function is an `async` function, as it'll return a promise.

Start our game with a `start()` function:

```
async function start() {
  const pokemon = await getPokemon()
}
```

We'll keep it simple. Note that this function also uses the async/await pattern and calls our function, which uses the API to retrieve a list of Pokémon. At this point, it would be a good idea to test our program by using `console.log()` to output the value of `pokemon`. You'll need to invoke the `start()` function in your program. You should see nice JSON data of 50 Pokémon.

In our `start()` function, we'll ask the player to choose their Pokémon with a message:

```
term.bold.cyan('Choose your Pokémon!\n')
```

Now we'll use our `pokemon` variable to create a grid menu with `term` to ask our player which Pokémon they'd like, as follows:

```
term.gridMenu(pokemon.map(mon => mon.name), {}, async (error, response) =>
{
  pokes['player'] = pokemon[response.selectedIndex]
  pokes['computer'] = pokemon[(Math.floor(Math.random() *
    pokemon.length))]
})
```

You can read the documentation on `term` to see more about what the options are for grid menus. We should run our code now, so in order to do that, add an invocation to the `start()` function at the end of the program:

```
start()
```

If we run our code with `npm start`, we'll see this new addition:

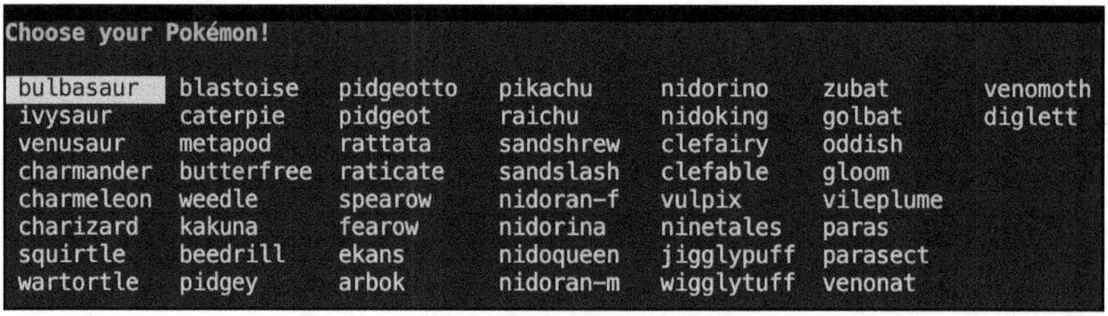

Figure 11.7 - Menu

With the arrow keys, we can navigate around the grid and choose our Pokémon by hitting *Enter*. In our code, we're assigning to our `pokes` object's two entries: `player` and `computer`. Now, `computer` will be a randomly selected entry from our `pokemon` variable.

We'll need more than the name and URL of our Pokémon to play, so we're going to make a helper function. Add this to our `start` function:

```
await createPokemon('player')
await createPokemon('computer')
```

Now we'll write the `createPokemon` function like so:

```
async function createPokemon(person) {
 let poke = pokes[person]

 const myPoke = await axios({
   url: poke.url,
   method: 'get'
 })
 poke = myPoke.data
 const moves = poke.moves.filter((move) => {
   const mymoves = move.version_group_details.filter((level) => {
     return level.level_learned_at === 1
   })
   return mymoves.length > 0
```

```
  })
  const move1 = await axios({
    url: moves[0].move.url
  })
  const move2 = await axios({
    url: moves[1].move.url
  })
  pokes[person] = {
    name: poke.name,
    hp: poke.stats[5].base_stat,
    img: await createImg(poke.sprites.front_default),
    moves: {
      [moves[0].move.name]: {
        name: moves[0].move.name,
        url: moves[0].move.url,
        power: move1.data.power
      },
      [moves[1].move.name]: {
        name: moves[1].move.name,
        url: moves[1].move.url,
        power: move2.data.power
      }
    }
  }
}
```

Let's unpack what this function is doing. First, we're going to get the information about our Pokémon (once for the player and once for the computer) from the API. The Pokémon moves section is a little more complicated, due to the fact that the gameplay is complex. For our purposes, we're simply going to assign the first two moves possible for our Pokémon in our `pokes` object.

For the image, we're using a small helper function:

```
async function createImg(url) {
  return asciify(url, { fit: 'box', width: 25 })
    .then((ascii) => {
      return ascii
    }).catch((err) => {
      console.error(err);
    });
}
```

We're almost done with the beginnings of our game! We need to add a few lines to our `gridMenu` method in `start`:

```
term.gridMenu(pokemon.map(mon => mon.name), {}, async (error, response) =>
{
    pokes['player'] = pokemon[response.selectedIndex]
    pokes['computer'] = pokemon[(Math.floor(Math.random() *
    pokemon.length))]
    await createPokemon('player')
    await createPokemon('computer')
    term(`Your ${pokes['player'].name} is so
    cute!\n${pokes['player'].img}\n`)
    term.singleLineMenu( ['Continue'], (error, response) => {
      term(`\nWould you like to continue against the computer's scary
      ${pokes['computer'].name}? \n ${pokes['computer'].img}\n`)
      term.singleLineMenu( ['Yes', 'No'], (error, response) => {
        term(`${pokes['computer'].name} is already attacking! No time to
        decide!`)
      })
    })
})
```

Now we can play!

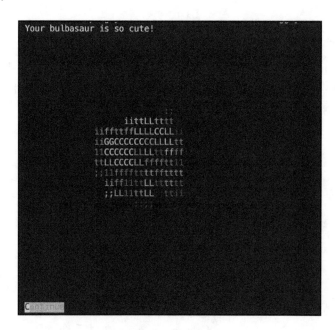

Figure 11.8 - Introducing your Pokémon!

The program continues with the computer's choice of Pokémon:

Figure 11.9 - The scary enemy Pokémon

For right now, we haven't included any actual gameplay using the moves and hitpoints. That can be a challenge for you to complete the `play()` function based on the logic from `Chapter 7`, *Events, Event-Driven Design, and APIs*.

The full code is here: `https://github.com/PacktPublishing/Hands-on-JavaScript-for-Python-Developers/tree/master/chapter-11/pokecli`.

Congratulations! We've done way more than "Hello, World!".

Summary

We've learned in this chapter that Node.js is a full-fledged programming language, capable of doing pretty much anything backend-related. We'll get into databases with Node.js in `Chapter 18`, *Node.js and MongoDB*, but, for the meantime, we can rest assured that it can do what we'd expect from a modern programming language.

The great thing about Node.js is that its grammar and structure *is* regular JavaScript! A few of the terms are different, but all in all, if you can read and write JavaScript, you can read and write Node.js. As with every language, there are differences in terminology and usage, but the fact is that Node.js and JavaScript are the same language!

In the next chapter, we'll discuss Node.js and Python and where certain choices make sense for using one versus the other.

Further reading

For more information, you can refer to the following:

- libuv: `https://en.wikipedia.org/wiki/Libuv`
- Market Share and Web Usage Statistics: `https://www.similartech.com/technologies/nodejs`

12
Node.js versus Python

Why would a developer choose Node.js over Python? Can they work together? What do our programs look like? These questions and more are at the heart of some of the differences between Python and Node.js, and it's important to understand when and where to use a particular language. For example, there are tasks for which a certain language is more suited than others, and it is the technologist's duty to advocate for the proper language. Let's investigate the use cases and different considerations when choosing Node.js versus Python.

The following topics will be covered in this chapter:

- Philosophical differences between Node.js and Python
- Performance implications

Technical requirements

Be prepared to work with the code provided in the `Chapter-12` directory of the repository at `https://github.com/PacktPublishing/Hands-on-JavaScript-for-Python-Developers/tree/master/chapter-12`. As we'll be working with command-line tools, also have your Terminal or command-line shell available. We'll need a modern browser and a local code editor.

Philosophical differences between Node.js and Python

It's common to have a main language that you know, work with, and are comfortable with. However, it's important to realize that not all programming languages are created for the same purpose. That is why it's very important to use the right tool for the job at hand. Just as you wouldn't attempt to build a house with a pocketknife, you probably wouldn't use a table saw to whittle a stick into a point for a campfire for s'mores.

If you've been in the industry for a while, you have probably heard the term **stack**. In technology, a stack is the architectural combination of technologies used to create a program or multiple programs in an ecosystem. In the past, applications tended to be large-scale **monoliths**, built in a "one application to rule them all" mindset. In today's world, the use of monoliths is decreasing in favor of multiple, smaller applications and **microservices**. In this manner, different parts of a workflow can be distributed to completely independent processes, greatly increasing the stability of an overall system.

Let's use Office software as an example. You certainly wouldn't try to write your next bestselling novel in Microsoft Excel, and you probably wouldn't want to do your taxes in Microsoft Word. There is a *separation of concerns* between these programs. They function very well together and form a unified whole, but each has its own part to play in your workflow.

Similarly, the different pieces of technology in a web application have their own uses and concerns. One of the more traditional stacks used for web applications is called **LAMP (Linux, Apache, MySQL, and PHP)**:

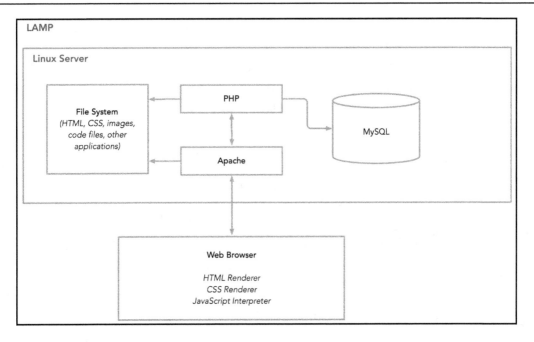

Figure 12.1 – LAMP stack

You can see that when discussing web applications in specifics, we take the web browser and the client stack as a given and unknown, not listed in the abbreviation LAMP. In this case, LAMP is only the server-side components.

As the web evolved, so did the underlying technologies supporting it and their stacks. Two of the more common stacks you may hear of now are **MEAN (MongoDB, Express, Angular, and Node.js)** and **MERN (MongoDB, Express, React, and Node.js)**. It's no coincidence that the only difference in technology is Angular versus React. They fulfill essentially the same role in an otherwise stable system. We'll explore Express, the ubiquitous web server framework for Node.js, in Chapter 13, *Using Express*, and MongoDB in Chapter 18, *Node.js and MongoDB*. For now, let's focus on the question of *why Node.js?*.

When choosing a language for a project, there are many factors to consider. Some of them are as follows:

- Type of project
- Budget
- Time-to-market
- Performance

These may sound like very basic factors, but I've certainly seen instances where the chosen technology was not a good fit for the type of project.

For those immersed in the web side of software, the choice between using JavaScript on the backend versus another language seems like a no-brainer. JavaScript is foundational to the use of the modern web, so it sounds like, by corollary, it should be used on both the client side and server side.

However, Python's been around longer and has definitely had a cemented foothold in the development community, especially with the explosion of interest in data science and machine learning, where Python reigns supreme. Flask and Django are excellent web frameworks that are robust and powerful. So why would we want to use Node.js instead?

The first part of deciding what tech stack to use is understanding the *type of project*. For the scope of our discussion today, let's limit our type of project to reasonable use cases. We won't open the can of worms to the Internet of Things/connected devices, as these are mostly written in Java. Let's also rule out machine learning and data science as possible use cases, as it's been established in the field that Python is better suited for these use cases. However, there actually is an argument for desktop and mobile applications being developed in JavaScript.

To begin with, let's think about if our project is a web application. In most cases, Node.js would be a logical fit over Python for many of the reasons that we've already explored: its asynchronous nature, less context switching, performance, and so forth. I'm hard-pressed to think of an adequate use case for using a Pythonic backend for a web app that would be superior to Node.js. I'm sure some exist, but in general, even when dealing with larger, more complex systems, the preference today is not to have a monolithic backend application, but rather to have a combination of *microservices* interacting with each other with data handoff.

Let's have a look at a possible **high-level architecture** (HLA) diagram. If you're dealing with complex applications, understanding the HLA of a system is extremely useful. Even if you're only actively working on one part of the application, understanding the needs and structure of other systems is invaluable. In this example, we have a possible architecture of an e-commerce site that also has a mobile app:

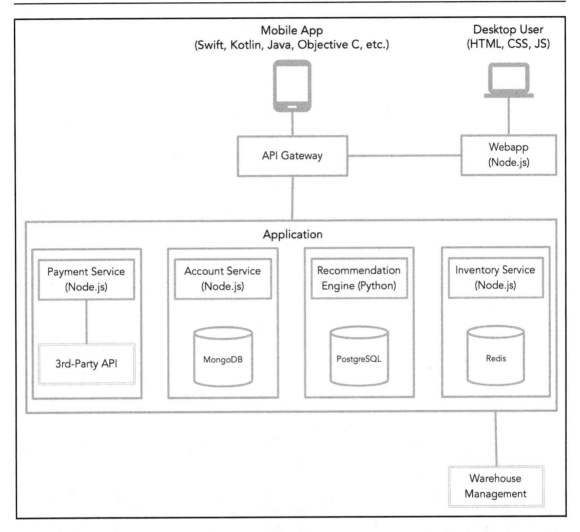

Figure 12.2 – High-level architecture

We can see that there can be multiple microservices, including some that are *not* Node.js or JavaScript at all. Python would be better suited as a microservice to provide the overarching application with recommendations, as that requires data analysis, which Python and R are better at doing than Node.js. Additionally, you can see that in the application, there can be multiple different data sources, from third-parties to different database types.

So, what about our project? Are we building this large ecosystem or a specific piece of it? In this example, the web app, payment service, account service, and inventory service are Node.js, as it makes sense to use technology designed for asynchronous communication. The recommendation engine, however, can be a *completely separate stack* without any problems, since it's contained within an overall ecosystem of microservices. As long as the parts of the application are appropriately communicating with each other, each service can be nearly independent.

Why is this important? Simply stated, it's a good way to enable smaller, nimbler teams to work in parallel to create software that is delivered faster and with greater stability than a monolithic application. Let's look at an example: you hit a major retailer's website to make a purchase, but instead of seeing the home page, you see the following:

500 Internal Server Error

cloudflare-nginx

Figure 12.3 – 500! Error, error, danger, danger!

The bane of any web application developer's life: a full-blown outage caused by a code problem. Instead, wouldn't it be much nicer if the site functioned for the majority of it, but, perhaps when it came time to check out, it said "Sorry, our payment processing system is currently offline. We've saved your cart for later." Or let's say the Pythonic piece of the recommendation engine crashed—we could instead serve in a static collection of items. To creatively construct an authentic user experience for a large ecosystem of microservices, it's important to consider the end user's standpoint *as well as* the business goals. In the case of our e-commerce store, we don't want the whole application to crash from a small error. Rather, we can intelligently downgrade the experience if problems were to occur. This is one example of a principle often known as fault-tolerant computing and, when designing large applications, it's powerful to separate a monolith into microservices in order to be more fault-tolerant.

I want to show you a quick example of some of the power of JavaScript existing in the desktop arena before we discuss the consideration of the budget. Let's run a piece of example code provided for you in the GitHub repository at `https://github.com/PacktPublishing/Hands-on-JavaScript-for-Python-Developers/tree/master/chapter-12/electron`:

1. Install the dependencies with `npm install`.
2. Run the application with `npm start`.

You should see a *native application* start—with our Pokémon game that we created in `Chapter 7`, *Events, Event-Driven Design, and APIs*:

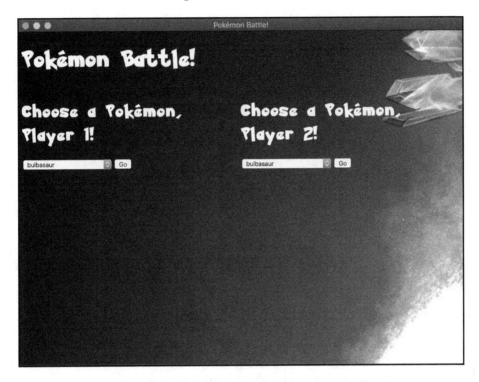

Figure 12.4 – It's a desktop app!

How did this happen? We leveraged a great tool: Electron. You can read more about Electron at `https://electronjs.org/`, but the gist is that it's a container tool to present HTML, CSS, and JavaScript as a desktop application. You may have already used Electron without realizing it: Spotify, Slack, and other popular desktop applications are built with Electron.

Let's take a quick look under the hood:

```
.
├── fonts
│   ├── pokemon_solid-webfont.woff
│   └── pokemon_solid-webfont.woff2
├── images
│   └── pokemon-2048x1152.jpg
├── index.html
├── main.js
```

```
├──── package-lock.json
├──── package.json
├──── poke.js
├──── preload.js
├──── renderer.js
└──── style.css
```

If we compare this to our PokéAPI project from Chapter 7, *Events, Event-Driven Design, and APIs*, (https://github.com/PacktPublishing/Hands-on-JavaScript-for-Python-Developers/tree/master/chapter-7/pokeapi/solution-code), we'll see that there are a lot of similarities.

Wait.

Not just similarities...this is *identical* to the code we used for our browser! main.js has been renamed to poke.js to avoid a naming conflict, but that's a small detail. Yes: you've just successfully created a desktop app with existing code.

So, back to budget: what if you need a web app *and* a desktop app? You should be getting the idea by now that with JavaScript, you can have the best of both worlds and have a modern web application *and* a desktop application with minimal changes. The nuances are a bit more than we've done here, but the power of Electron should be evident. Write once, use multiple times—isn't that the mantra of DRY coding?

However, there is a flip side to this argument. Since Python has been mature longer than Node.js, there is a probability that Python developers will be more cost-effective in their hourly rate than Node.js developers. However, I would consider this a secondary concern.

Likewise, as a secondary concern, *time-to-market* is indeed a question that arises when choosing a technology. Unfortunately, the numbers here are inconclusive. Because Node.js is JavaScript, in theory, it can be developed quickly and iteratively. However, Python's explicit and simple syntax makes writing it faster at times. It's a very difficult problem to solve, so it's best to consider another part of the timing aspect: technical debt. Tech debt is the bane of engineering teams, and it simply means that, at the expense of the optimal solution, a faster solution was implemented. Additionally, the attrition of technology can lead to tech debt. Do you remember Y2K? When it was discovered that many major applications in the world were dependent on a two-digit year, it was feared that the change from the year 1999 to 2000 would wreak havoc on computer systems. Thankfully, only minor glitches occurred, but the problem of tech debt arose: many of those systems were coded in languages that had since become obscure. Finding programmers to develop these fixes was difficult and costly. Likewise, if you choose a technology because it's faster, you may find yourself paying twice or thrice the initial investment in terms of budget and time to refactor the application to meet ongoing needs.

Let's turn our attention to performance. There is a lot to consider here, so let's continue to the next section and discuss why performance is always a consideration when discussing Node.js.

Performance implications

When Node.js was first gaining popularity, there were concerns about its single-threaded nature. Single thread means one CPU, and one CPU can be overwhelmed by large traffic influxes. However, for the most part, all of these thread concerns have been assuaged by advances in server technology, hosting, and DevOps tools. With that being said, the single-threaded nature also shouldn't be a deterrent in and of itself: we'll discuss in just a little bit why the *Node event loop* plays an important role in any discussion around Node.js's performance.

In a nutshell, to really differentiate on performance, we should focus on *perceived* performance. Python is an easy-to-understand, robust, object-oriented programming language; there are no disputes about that. However, one of the things it does not, cannot, and will not do is run in the browser. That spot is taken by JavaScript.

Why is this important and how does it relate to performance? In a nutshell: *Python can't react to changes in the browser.* It's possible to perform an Ajax request every time the UI of a page changes, but that would be incredibly expensive computationally both for the browser and the server. Additionally, you'd have to make the browser wait for a response from the server at each change, causing a very laggy experience. So, therefore, the more we can do client-side in the browser, the better. Using JavaScript in a browser to handle logic before needing to communicate with a server is the goal.

Implicit in the discussion around using Node.js is a thought you probably have from the previous section: *Node.js doesn't run in the browser either!* That's true! However, there's the fact that Node.js is based on the Chrome interpreter and, as such, implicit in its design is the idea of asynchronicity. The event loop of Node.js is designed for, well, events—and inherent in the concept of events is that they're asynchronous.

Let's review the following diagram from `Chapter 7`, *Events, Event-Driven Design, and APIs*:

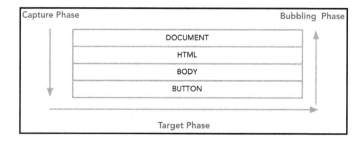

Figure 12.5 – The event life cycle

If you recall, this diagram represents the three phases of a browser event: capturing, targeting, and bubbling. The DOM events specifically relate to actions, interactions, or triggers that are caused in the browser either by the user or the program itself.

Similarly, the event loop of Node.js has a life cycle:

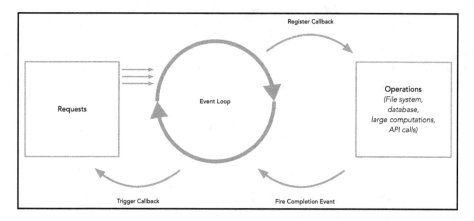

Figure 12.6 - The Node.js event loop

Let's decipher this. The single-thread event loop runs during the life cycle of a Node application and accepts inbound requests, either from a browser, another API, or other sources, and executes its work. If it's a simple request or specified as synchronous, it can be returned immediately. For more intensive operations, Node will register a *callback*. Remember that this is a term for a function that is passed to another function in order to be executed when the function it's passed to finishes its work. We've used these extensively as *event handlers* in JavaScript so far. The Node.js event loop provides an efficient means to access and serve data to our application.

If you're not too familiar with the concepts of threads and processes, that's OK because we won't really dive in deep here. However, it's important to point out some facts about Node's use of processes and threads. It's stated by some computer scientists that the single-threaded nature of Node is inherently unscalable and that it can't stand up to the volume of traffic that a mature web app would need. However, as I mentioned previously, our applications don't live in a silo. No application that needs to be designed for scale is just going to hang out on a server alone. With the advent of cloud technologies such as Amazon AWS, it's easy to incorporate multiple virtual machines, load balancers, and other virtual tools to appropriately distribute the load for an application. Yes, Python may be better suited as a single-box application to receive thousands upon thousands of incoming requests, but this benchmark for performance is outdated and belies the state of technology today.

Caveat emptor

Now that we're in love with Node, let's return to the idea of the right tool for the task at hand. Node isn't a panacea for all the world's computing problems. In fact, it's specifically *not* designed to be a Swiss Army knife. It has its use and its place, but it doesn't try to be everything to everyone. The "do everything" nature of Java might be considered one of its weaknesses, in that while you can write Java code once and compile it for practically any architecture, there are limitations, considerations, and trade-offs that have been made to accommodate this. Node.js and JavaScript, in essence, try to stay in their own lanes.

So, what's the catch? We know that JavaScript is fast, powerful, effective, and understandable. Like any technology, there are always nuances, and one nuance of JavaScript and Node is this mantra that is in some Linux systems when you first log in as a superuser: "With great power comes great responsibility." While the provenance of the quote is nebulous, it's important to think of when executing anything that has influence over others. (Don't use hypnotism for evil!)

All joking aside, there are very real problems that can occur with an asynchronous environment. We know that we can easily crash a user's browser with our own client-side JavaScript code by throwing it into an infinite loop. Consider the following code:

```
let text = ''

while (1) {
  text += '1'
}
```

Excellent. If you were to run this in a browser, the *best* that could happen is that the browser would recognize an infinite loop and prompt a warning for you to exit the script because the page is unresponsive. A second outcome would be that the browser crashes, and in the worst-case scenario, the user's entire machine may crash with an out-of-memory exception. With great power...

Similarly, we can severely impact the user's experience in Node by improperly handling state and events. What if, for example, your frontend code was dependent upon a Node process and that process never returned? Luckily, there are Ajax safeguards built in to prevent this in most cases, in that HTTP requests will, by default, close after a certain period of time and error out if necessary. With that being said, there are a whole host of ways to force a connection to stay open, causing absolute chaos to a user's browser. There are perfectly valid reasons for doing this, such as long polling for live data, so that's why they exist. On the flip side, it's also possible to accidentally cause major issues for a user. Fail safes such as timing out requests exist for you protection, but any good engineer will tell you: don't rely on fail safes—avoid errors in the design process.

Summary

Python is great. Node is great. Both are great. So why are we having this conversation? While both technologies are strong and mature, each has a role to play in the technology ecosystem. Not all languages are created equal, and not all languages handle problems in the same way.

To summarize, we have learned the following:

- Node.js is asynchronous and works well with event-based thoughts, such as JavaScript in a browser reacting to page events.
- Python has established itself as a leader in data analysis and machine learning, as it's able to process large datasets quickly.
- For web work, the technologies may be interchangeable, but a complex architecture may involve both (and more!).

In the next chapter, we'll begin working with Express, a foundational web server for Node.js. We'll create our own websites and work with them.

Further reading

Here's a little more reading on some of these topics:

- stateofjs: `https://2019.stateofjs.com/`
- NodeJs versus Python: `https://www.similartech.com/compare/nodejs-vs-python`
- Pattern — microservice architecture: `https://microservices.io/patterns/microservices.html`
- Amazon API Gateway: `https://aws.amazon.com/api-gateway/`
- Electron: `https://electronjs.org/`
- Y2K bug: `https://www.britannica.com/technology/Y2K-bug`
- Node.js multithreading: `https://blog.logrocket.com/node-js-multithreading-what-are-worker-threads-and-why-do-they-matter-48ab102f8b10/`

13
Using Express

As we've discussed, JavaScript in the backend is incredibly useful for creating web applications and harnessing the power of JavaScript on both the front- and back-end. One of the most fundamental tools for a server-side application that interacts with the frontend is a basic web server. In order to provide APIs, database access, and other functionality that is not designed to be handled by the browser, we first need to set up a piece of software to handle these interactions.

Express.js (or just Express) is a web application framework, considered the *de facto* web server of Node.js. It enjoys both high popularity and ease of use. Let's use it to build a full-blown web app.

The following topics will be covered in this chapter:

- The scaffold: Using `express-generator`
- Routes and views
- Controllers and data: Using APIs in Express
- Creating an API with Express

Technical requirements

Be prepared to work with the GitHub repository at `https://github.com/PacktPublishing/Hands-on-JavaScript-for-Python-Developers/tree/master/chapter-13`. All you'll need is a code editor and a browser. In the *Routes and views* section, we'll discuss a few best practices for working with code editors.

The command-line examples are presented in macOS/Linux style. Windows users might need to consult the documentation to understand a few nuances of the Windows command line.

The scaffold: Using express-generator

To get started, we'll need to get on our **command-line interface** (**CLI**) again. If you remember Chapter 2, *Can We Use JavaScript Server-Side? Sure!*, we took a look at Node and npm on the command line. Let's check our version again so we can make a few decisions about our application. On your command line, run node -v. If you have v8.2.0 or greater, you have the option of using npx to install certain packages that are designed to be run only once in the lifespan of a project, such as express-generator. However, if you have a lower version, you can use npm to install one-time-use packages as well as packages that are used in your project.

We'll move forward with npx in this chapter, so if you need to take a quick look at the documentation for npm versus npx, be sure to give yourself some time to do that. In essence, to use npm for one-time packages that shouldn't live inside your code base, for example, a scaffolding tool such as an Express generator or a React app creator, you can install the package globally on your system like this: npm install -g express-generator. Then, you'll run the program with Express. However, this is considered a legacy usage of npm as npx is favored in today's landscape.

Let's create our Express app from scratch to build up muscle memory instead of continuing on from Chapter 2, *Can We Use JavaScript Server-Side? Sure!*. Follow these steps to get started on an Express web server:

1. In a convenient location, create a new directory with mkdir my-webapp.
2. Navigate inside it with cd my-webapp.
3. The npx express-generator --view=hbs express generator will create several files and directories:

```
→   my-webapp npx express-generator --view=hbs
npx: installed 10 in 5.254s

    create : public/
    create : public/javascripts/
    create : public/images/
    create : public/stylesheets/
    create : public/stylesheets/style.css
    create : routes/
    create : routes/index.js
    create : routes/users.js
    create : views/
    create : views/error.hbs
    create : views/index.hbs
    create : views/layout.hbs
    create : app.js
    create : package.json
    create : bin/
    create : bin/www

    install dependencies:
      $ npm install

    run the app:
      $ DEBUG=my-webapp:* npm start
```

Figure 13.1 - Creating our Express scaffold

We're going to set our application to use Handlebars for our templating instead of Jade, which is the default option. Express supports multiple templating languages out of the box (and can be extended to use any), but for ease of use, we'll stick with Handlebars, which is very similar to how we worked with frontend frameworks such as React and Vue in Chapter 8, *Working with Frameworks and Libraries*, with basic curly brace tokens.

4. Use npm install to install our dependencies. (Note that even if you used npx before, you're going to use npm here.) This is going to take a few seconds and will download a number of packages and other dependencies. Another thing to note is that you will need an internet connection, as npm retrieves packages from the internet.

5. Now, we're ready to start our application using `npm start`:

```
➜ my-webapp git:(master) ✗ npm start
pm audit fix to fix them, or npm audit for details

➜ my-webapp@0.0.0 start /Users/sonylnagale/Documents/Hands-on-JavaScript-for-Python-Developers/chapter-13/my-webapp
➜ node ./bin/www
```

Figure 13.2 - Our application starting

6. OK! Now, let's access our Express site in a web browser:

Figure 13.3 - Express welcome page

Fantastic! Now that we're at this step, let's take this further than we did in `Chapter 2`, *Can We Use JavaScript Server-Side? Sure!*.

RESTful architecture

At the core of many web apps is a REST (or RESTful) application. **REST** is an abbreviation of **REpresentational State Transfer**, which is a design pattern that deals with the fact that most web technologies are inherently **stateless**. Think of a standard website that doesn't require a login or much data—just static HTML and CSS, as we created in the previous chapters, but even simpler: without JavaScript. If we think of a site like this in terms of state, we can see that a bunch of HTML doesn't know our user journey, doesn't know who we are, and, quite frankly, it doesn't care. Sites like these are like printed material: you interact with it by looking, reading, and turning pages. You don't change anything about it. In general, the only way you're really modifying the state of a book is with a bookmark to save your place. Honestly, that's one step more interactive than a basic blob of HTML and CSS!

To work with users and data, REST is used as a functional paradigm. We already worked with two of the main HTTP verbs when working with APIs: GET and POST. These are the two main verbs that we'll be using, but we'll be looking at two more: PUT and DELETE.

If you're familiar with the concept of **Create, Read, Update, and Delete (CRUD)**, this is how the standard HTTP REST verbs translate:

Concept	HTTP Verb
Create	CREATE
Read	GET
Update	PUT or PATCH
Delete	DELETE

 For more information, you can take a look at the Packt REST Tutorial: `https://hub.packtpub.com/what-are-rest-verbs-and-status-codes-tutorial/`.

Now, it is possible to create a full application only using GET, or just GET and POST, but there are security and architectural reasons why you wouldn't want to do this. For now, let's agree to follow best practices and work within this established paradigm.

Now, we're going to create a RESTful application.

Routes and views

Routes and views are the foundation of how a RESTful application's URLs act as pathways to its logic and how content is presented back to the user. Routes will determine what parts of code correspond to the URLs of the application's interface. Views determine what is displayed, either to a browser, another API, or other programmatic access.

To further understand the structure of an Express application, we can examine its routes and views:

1. First of all, let's open our Express application in your favorite IDE. I'm going to be working with VS Code. If you use VS Code, Atom, Sublime, or another IDE that has command-line tools, I highly recommend installing them. For example, with Atom, you can launch a multi-panel Atom editing interface by typing `atom .` in the command prompt and opening that directory in Atom.

2. Similarly, VS Code will do this with `code` .. Here's what this looks like:

Figure 13.4 - VS Code

I've expanded the directories on the left so that we can see the first level of the hierarchy.

3. Open `app.js`.

One of the first things you'll notice is that the syntax of this code that express-generator created for us is *ES5*, not ES6. For the moment, let's not concern ourselves with converting it to ES6; we'll do that a bit later. As we work on our first Node.js REST application, keep in mind that there are a couple of different ways of achieving our goal, and we'll take the more verbose path first in order to get our functionality working and then iterate on it to make it more flexible and more DRY.

4. Right now, you shouldn't have to make any changes to `app.js`, but do take a second to familiarize yourself with its structure. One of the more unfamiliar aspects of it may be the `require()` statements at the beginning of the file. Similar to how you have used `import` in a frontend framework, `require()` is Node's way of signaling to its component system to bring in these pieces from other files. In this case, the first few lines are bringing in modules installed via `npm`, like so:

```
var express = require('express');
```

Note that there is no path in front of `('express')`. It's simply stated. This is an indicator that the module referenced is not native to our code. If you look at the `require` statement for `indexRouter`, however, we see that it *does* have a path: `'./routes/index'`. It does *not* have the `.js` extension, but it's pathed properly for our module usage.

Now, let's examine our `routes/index.js` file.

Routes

If you open `routes/index.js`, you'll see the following few lines of code that were generated for us:

```
var express = require('express');
var router = express.Router();

/* GET home page. */
router.get('/', function(req, res, next) {
    res.render('index', { title: 'Express' });
});

module.exports = router;
```

There isn't a whole lot surprising here: as we're starting to gather, Express files begin with `require` statements, especially for `express` itself. In the next code block, we're starting to see the beginnings of our REST service: GET home page. Look at the `router.get()` method right after the comment. It's *explicitly* stating to the router that when a GET request is received for the URL of `/`, execute this code.

We can verify this fact by adding a few more GET paths here, just for fun. Let's try modifying our code like so. After the `router.get()` block, but before `module.exports`, let's register more routes on the router:

```
/* GET sub page. */
  router.get('/hello', function(req, res, next) {
      res.render('index', { title: 'Hello! This is a route!' });
  });
```

Now, we must stop our Express server with *Ctrl + C*, restart it with `npm start`, and access our new page at `http://localhost:3000/hello`:

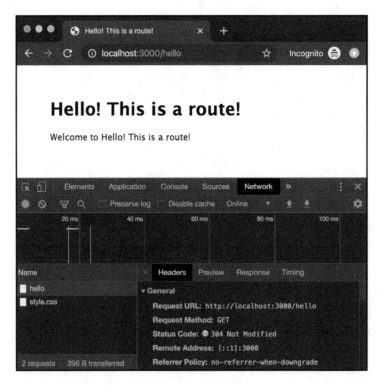

Figure 13.5 - A new route, with the Network tab open, showing we are making a GET request

So far, this should seem pretty basic. Now, let's do something a little different. Let's work with this view and create a form for an Ajax POST request:

1. Create a new file called `public/javascripts/index.js`.
2. Write a basic `fetch` request to the endpoint `/hello` with POST JSON of `{ message: "This is from Ajax" }` like so:

```
fetch('/hello', {
 method: 'POST',
 body: JSON.stringify({ message: "This is from AJAX" }),
 headers: {
   'Content-Type': 'application/json'
 },
});
```

3. Include this file in `views/index.hbs` like so:

```
<h1>{{title}}</h1>

<p>Welcome to {{title}}</p>

<p id="data">{{ data }}</p>

<script src="/javascripts/index.js"></script>
```

   Notice that we don't need to include `public` in our path. This is because Express already understands that files within `public` are to be served statically without any intervention or parsing from Express, as opposed to Node files that must be run.

4. If you reload the page now, you really won't see anything exciting happening because we haven't yet written the route to handle the POST request. Write it like this:

```
/* POST to sub page. */
router.post('/hello', function(req, res, next) {
  res.send(req.body);
});
```

5. Reload your page and you'll see... nothing. No POST in the **Network** tab, and certainly nothing rendered. What happened?

Node has several tools that are used to reboot Express servers when code is changed so that the engine will refresh itself without us needing to kill and restart it, as we were doing before, but we did not do this time. These tools change over time, but the one I like is Supervisor: `https://www.npmjs.com/package/supervisor`. Install it in your project simply by executing `npm install supervisor` in the directory of your project.

6. Now, open the `package.json` file in the root of your project. You should see a file similar to this, but perhaps with some version differences:

```
{
  "name": "my-webapp",
  "version": "0.0.0",
  "private": true,
  "scripts": {
  "start": "node ./bin/www"
  },
  "dependencies": {
  "cookie-parser": "~1.4.4",
  "debug": "~2.6.9",
  "express": "~4.16.1",
  "hbs": "~4.0.4",
  "http-errors": "~1.6.3",
  "morgan": "~1.9.1",
  "supervisor": "^0.12.0"
  }
}
```

This is the core of what is installed when running `npm install`. When you run it, you'll see a `node_modules` directory created and many files written there.

 If you're using version control such as Git, you will *not* want to commit the `node_modules` directory. With Git, you would include `node_modules` in a `.gitignore` file.

7. The next thing we want to do is alter our start script to now use Supervisor:

```
"scripts": {
    "start": "supervisor ./bin/www"
},
```

To use it, we still use `npm start` and to quit it, you just press *Ctrl + C*. It's worth noting that Supervisor is best for local development work, not production work; there are other tools such as Forever for that purpose.

8. Now, let's run `npm start` and see what happens. You should see some console messages that end with **Press rs for restarting the process**. Under most circumstances, issuing `rs` is not needed, but it is there if you need it:

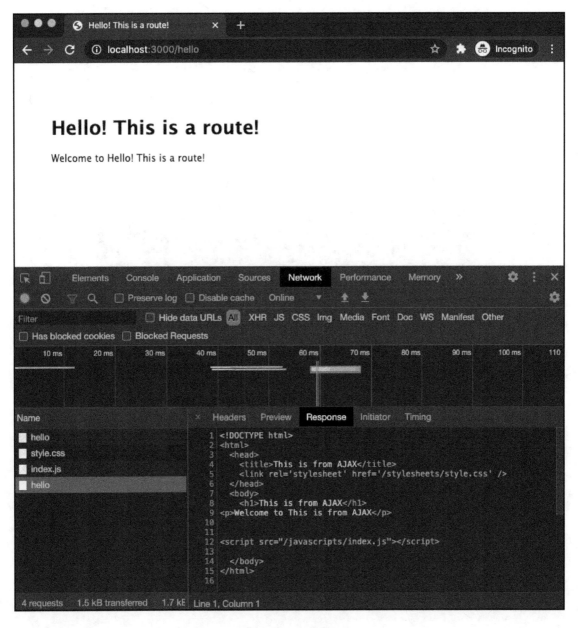

Figure 13.6 - Response from Ajax!

9. Since we sent `This is from AJAX` from our frontend JavaScript, we're seeing it reflected in our response HTML! Now, if we were to want it in our page, we would do this in our frontend JavaScript:

```
fetch('/hello', {
 method: 'POST',
 body: JSON.stringify({ message: "This is from AJAX" }),
 headers: {
    'Content-Type': 'application/json'
 },
}).then((res) => {
 return res.json();
}).then((data) => {
 document.querySelector('#data').innerHTML = data.message
});
```

We'll see the following:

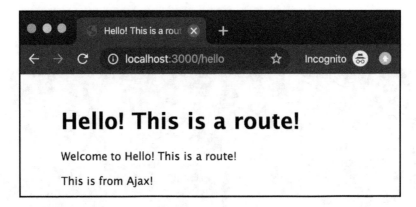

Figure 13.7 - A message from Ajax!

Next, let's understand how to save data.

Saving data

For our next step, we'll persist data in a local data store, which will be a simple local JSON file:

1. Go ahead and quit Express with *Ctrl* + *C*. Let's install an easy module that saves data in a local store: `npm install data-store`.

2. Let's modify our routes to use it, like so:

```
var express = require('express');
var router = express.Router();

const store = require('data-store')({ path: process.cwd() +
'/data.json' });

/* GET home page. */
router.get('/', function(req, res, next) {
 res.render('index', { title: 'Express', data:
 JSON.stringify(store.get()) });
});

/* GET sub page. */
router.get('/hello', function(req, res, next) {
 res.render('index', { title: 'Hello! This is a route!' });
});

/* POST to sub page. */
router.post('/hello', function(req, res) {
 store.set('message', { message: `${req.body.message} at
${Date.now()}` })

 res.set('Content-Type', 'application/json');
 res.send(req.body);
});

module.exports = router;
```

3. Notice the inclusion of `store` and its use in the `hello` and `/` routes. Let's also modify our `index.hbs` file like this:

```
<h1>{{title}}</h1>
<p>Welcome to {{title}}</p>

<button id="add">Add Data</button>
<button id="delete">Delete Data</button>

<p id="data">{{ data }}</p>
<script src="/javascripts/index.js"></script>
```

4. We'll use the `Delete Data` button later, but for now, we'll work with the `Add Data` button. Add some save logic to `public/javascripts/index.js` like so:

```
const addData = () => {
 fetch('/hello', {
```

```
      method: 'POST',
      headers: {
        'Content-Type': 'application/json'
      },
      body: JSON.stringify({ message: "This is from Ajax" })
  }).then((res) => {
    return res.json()
  }).then((data) => {
      document.querySelector('#data').innerHTML = data.message
  })
}
```

5. And now we'll add our click handler:

```
document.querySelector('#add').addEventListener('click', () => {
  addData()
  window.location = "/"
})
```

6. If you refresh the / page and click the **Add Data** button, you should see something like this:

Figure 13.8 - Adding data

7. Now, refresh that page again. Notice that the message persists. In your filesystem, you should also notice a `data.json` file that contains the data.

We're now ready to work with this a bit more with a delete method.

Delete

We've explored GET and POST, and now it's time to deal with another foundational REST verb: **DELETE**.

As its name implies, its goal is to remove data from a data store. We already have our button to do so, so let's wire it up:

1. In our frontend JavaScript, we'll add the following:

```
const deleteData = () => {
  fetch('/', {
    method: 'DELETE',
    headers: {
      'Content-Type': 'application/json'
    },
    body: JSON.stringify({ id: 'message' })
  })
}
document.querySelector('#delete').addEventListener('click', () => {
  deleteData()
  window.location = "/"
})
```

2. Now, in the routes, add this route:

```
/* DELETE from json and return to home page */
router.delete('/', function(req, res) {
  store.del(req.body.id);

  res.sendStatus(200);
});
```

And that should be all we need. Refresh your index page and play around with the add and delete buttons. Pretty easy, right? In Chapter 18, *Node.js and MongoDB*, we'll discuss persisting and manipulating our data in a full-fledged database, but for now, we can work with the knowledge of GET, POST, and DELETE. We'll work with PUT with an actual database.

Views

We touched on the manipulation of views in the *Routers* section, so let's now dive in deeper. The **view layer** of an application is the presentation layer, which is why it houses our frontend JavaScript. While not all backend Node applications will serve a frontend, it's handy to know how to use it. Whenever I set up a simple webserver, I reach for Express and its functionality for both the frontend and backend.

Since there are multiple frontend templating languages available to us, let's use Handlebars as an example of logic and structure.

If we want to, we can provide some conditional logic in our frontend code. Note that this logic is rendered by the backend, so it's a good example of when to render data on the backend (which is more performant for the frontend) and when to do it via JavaScript (which is, ostensibly, more flexible).

Let's alter our `views/index.hbs` file like this:

```
{{#if data }}
  <p id="data">{{ data }}</p>
{{/if}}
```

Let's also modify `routes/index.js`:

```
/* GET home page. */
router.get('/', function(req, res, next) {
  res.render('index', { title: 'Express', data:
  JSON.stringify(Object.entries(store.get()).length > 0 ? store.get() :
  null) });
});
```

Now, we're using a ternary operator to simplify our display logic. Since our data from the store is JSON, we can't simply test its length: we have to use the `Object.entries` method. If you're thinking that we could have saved the `store.get()` to a variable instead of writing it twice, you're right. However, in this case, we don't really need to take up additional memory space since we're immediately returning it versus manipulating it. The performance impact in this scenario is negligible, but it is something to keep in mind.

Now, if we delete our data, we will see this:

Figure 13.9 - After deleting data

It's a little less confusing to see than to see an empty object's curly braces. Sure, we could have done the conditional work on the frontend by writing a more complex conditional, but why do that work when the backend can send us the appropriate data? There are certainly cases for both, but in this case, it's better to let each piece do its own work.

You can find our completed work here: `https://github.com/PacktPublishing/Hands-on-JavaScript-for-Python-Developers/tree/master/chapter-13/my-webapp`.

Let's now turn our attention to how we actually get data into Express using **controllers**.

Controllers and data: Using APIs in Express

As you may have heard around the web, Express is great because it's not very opinionated on how you use it, and at the same time, people say that Express is hard to work with because it's not opinionated enough! While Express isn't typically set up as a traditional Model-View-Controller setup, it can be beneficial to split functionality out of your routes and into separate controllers, especially if you may end up having similar functionality between routes and want to keep your code DRY.

If you're not very familiar with the **Model-View-Controller** (**MVC**) paradigm, don't worry—we won't go into it in too much detail, as it's a very weighty topic, complete with its own debates and conventions. For now, we'll just define a few terms:

- A **Model** is a part of the application that deals with data manipulation, especially communication to and from a database.
- A **Controller** deals with logic from routes (that is, the path of the HTTP request from the user).
- The **Views** are the presentation layer to the end-client that provides markup to the client, routed by the controller.

This is what the MVC paradigm looks like:

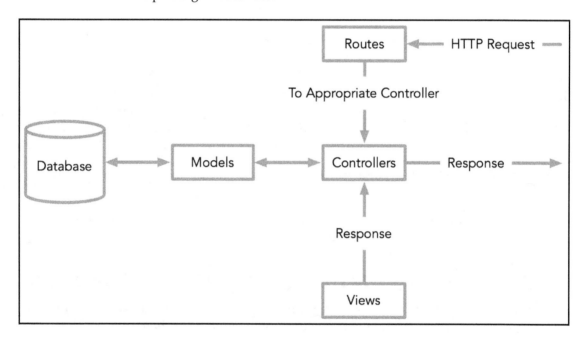

Figure 13.10 - The MVC paradigm

Let's take a look at a sample application. At `https://github.com/PacktPublishing/Hands-on-JavaScript-for-Python-Developers/tree/master/chapter-13/controllers` is an application that uses Express.

This is an API that uses controllers and models. As we'll see, this structure will simplify our workflow. It's still a fairly simple example, but this will give you a sense of why controllers and models can come in handy. Let's investigate it:

1. Go ahead and run `npm install`, and then run `npm start` to run the application. It should be accessible in your browser at `http://localhost:3000`, but if you have anything else running, Node will warn you and state a different port. Here's what you'll see:

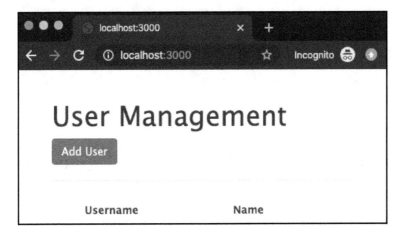

Figure 13.11 - Our sample Express application

2. Pretty simple so far. Go ahead and click **Add User** a few times and play around with the functionality. This uses a random user API on the backend to create users and persist them to a filesystem data store.
3. Examine the client-side JavaScript in the `public/javascripts` directory. This should all look familiar. If we remember the structure of the `fetch()` call, it returns a promise, so we can use the `.then()` paradigm to react to our events.
4. In `public/javascripts/index.js`, we can see the mechanism that creates our users when we click **Add User**:

```
document.querySelector('.add-user').addEventListener('click', (e)
=> {
  fetch('/user', {
    method: 'POST'
  }).then( (data) => {
    window.location.reload()
  })
})
```

This shouldn't be anything surprising: we're using JavaScript's `fetch` in an event handler to call the `/user` route with a POST. A **route** is basically an endpoint in an Express (or other) application: it contains some logic to react to an event. So, what is that logic?

5. Open `routes/user.js`:

```
var express = require('express');
var router = express.Router();

const UsersController = require('../controllers/users');

/* GET all users. */
router.get('/', async (req, res, next) => {
  res.send(await UsersController.getUsers());
});

/* GET user. */
router.get('/:user', async (req, res, next) => {
  const user = await UsersController.getUser(req.params.user);
  res.render('user', { user: user });
});

/* POST to create user. */
router.post('/', async (req, res, next) => {
  await UsersController.createUser();
  res.send(await UsersController.getUsers());
});

/* DELETE user. */
router.delete('/:user', async (req, res, next) => {
  await UsersController.deleteUser(req.params.user);
  res.sendStatus(200);
});

module.exports = router;
```

Let's first compare the structure of this to the other examples. First, we'll see the `require()` statement for the controller for users. There's a `router.post()` method statement here, which is using `async/await` for asynchronous calls to our controller. Our controller will then call our model to do database work.

So far, there are a good number of files and paths for execution. Before we get too lost in the code, let's take a look at a diagram of how a frontend method, such as the **Add User** click handler, communicates with our Express backend:

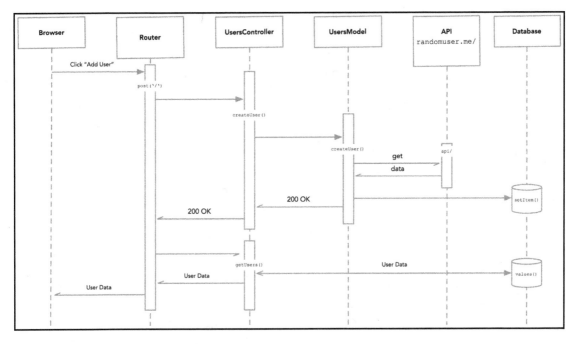

Figure 13.14 - End-to-end communication

Reading from left-to-right and top-to-bottom, we can see how each step in the process plays its role. It may *look* a bit complicated for something as basic as retrieving information from an API, but part of the power of this architectural pattern is that each layer can be written and controlled by a different party. For example, the model layer will often be in the hands of database experts, as opposed to other types of backend developers.

As you trace through the code for the controller and model, consider how the separation of concerns at each layer of the code makes for a more modular design. For example, we're using a LocalStorage database to store our users. If we wanted to swap LocalStorage for a more robust system, such as MongoDB, we would really only have one file to edit: the model. In fact, even the model may be abstracted to have unified data handlers and then use an adapter for specific database methods.

That might be a bit much for us right now, but let's next turn our sights on creating a starship game using the principles we just learned. We'll use this Node.js backend for our final project to make a frontend for the game in JavaScript.

In the next section, we'll get started with creating our game's API.

Creating an API with Express

Who doesn't like a nice starship battle, like in Star Wars or Star Trek? I happen to be quite a fan of science fiction, so let's play along and construct a RESTful API using storage, routes, controllers, and models to keep track of our gameplay. While we'll be focusing on the backend of this application, we'll stand up a simple frontend for the population of data and testing.

You can find a work-in-progress example app at https://github.com/PacktPublishing/ Hands-on-JavaScript-for-Python-Developers/tree/master/chapter-13/starship-app. Let's start there, and you can finish it using the following steps:

1. Clone the repository if you haven't already.
2. Navigate into the directory with cd starship-app and run npm install.
3. Start the project with npm start.
4. Open http://localhost:3000 in a browser. If you already have any projects running on port 3000, the start command may prompt you to use a different port. Here's our basic frontend:

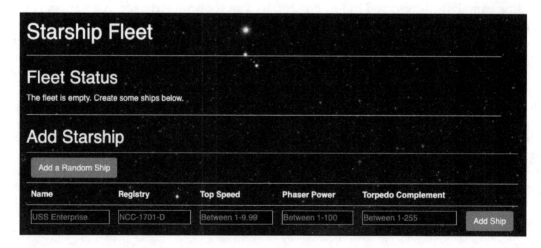

Figure 13.15 - Starship Fleet

5. Go ahead and play around with adding and scuttling ships both randomly and manually. This will be the setup for our gameplay.

6. Now, let's unpack what the code is doing. Here's our file structure:

```
.
├──── README.md
├──── app.js
├──── bin
│    └──── www
├──── controllers
│    └──── ships.js
├──── data
│    └──── starship-names.json
├──── models
│    └──── ships.js
├──── package-lock.json
├──── package.json
├──── public
│    ├──── images
│    │    └──── bg.jpg
│    ├──── javascripts
│    │    └──── index.js
│    └──── stylesheets
│         └──── style.css
├──── routes
│    ├──── index.js
│    ├──── ships.js
│    └──── users.js
└──── views
     ├──── error.hbs
     ├──── index.hbs
     └──── layout.hbs
```

7. Open `public/javascripts/index.js`. Let's first examine the event handler for random ship creation:

```
document.querySelector('.random').addEventListener('click', () => {
  fetch('/ships/random', {
    method: 'POST'
  }).then( () => {
    window.location.reload();
  })
})
```

So far so good. This should all look familiar.

8. Let's examine this route: /ships/random. Open routes/ships.js (we can guess that the routing for /ships/ will be in the ships.js file, but we can confirm this by reading the routing in the app.js file, as we've learned). Read through the /random route:

```
router.post('/random', async (req, res, next) => {
  await ShipsController.createRandom();
  res.sendStatus(200);
});
```

The first thing we'll notice is that this is an async/await construction, as we're going to be working with fetch on the frontend and (spoiler alert) a database on the backend.

9. Let's look at that controller method next:

```
exports.createRandom = async () => {
  return await ShipsModel.createRandom();
}
```

10. Easy enough. Now for the model method:

```
exports.createRandom = async () => {
  const shipNames = require('../data/starship-names');
  const randomSeed = Math.ceil(Math.random() *
  shipNames.names.length);

  const shipData = {
    name: shipNames.names[randomSeed],
    registry: `NCC-${Math.round(Math.random()*10000)}`,
    shields: 100,
    torpedoes: Math.round(Math.random()*255+1),
    hull: 0,
    speed: (Math.random()*9+1).toPrecision(2),
    phasers: Math.round(Math.random()*100+1),
    x: 0,
    y: 0,
    z: 0
  };

  if (storage.getItem(shipData.registry) || storage.values('name')
  == shipData.name) {
    shipData.registry = `NCC-${Math.round(Math.random()*10000)}`;
    shipData.name = shipNames.names[Math.round(Math.random()*
    shipNames.names.length)];
```

```
        }
        await storage.setItem(shipData.registry, shipData);
        return;
    }
```

OK, it's a bit more complicated, so let's unpack this. The first couple of lines are simply selecting a random name from a seed file provided for you. Our `shipData` object is constructed from several key/value pairs, each corresponding to specific properties of our newly-created ship. After that, we check our database to see whether there's already a ship by that name or registry number. If so, we'll randomize again.

However, as with every application, there are areas for improvement. Here's a challenge for you.

Challenge

Here's your first task: can you think of how to improve the code so that in the re-randomization, it elegantly checks to see whether the *new* randomization also exists in our database? Hint: you may want to create a separate helper function or two.

Maybe you arrived at something like this (`https://github.com/PacktPublishing/Hands-on-JavaScript-for-Python-Developers/tree/master/chapter-13/starship-app-solution1`):

```
const eliminateExistingShips = async () => {
  const shipNames = require('../data/starship-names');
  const ships = await storage.values();

  const names = Object.values(ships).map((value, index, arr) => {
    return value.name;
  });

  const availableNames = shipNames.names.filter((val) => {
    return !names.includes(val);
  });

  const unavailableRegistryNumbers = Object.values(ships).map((value, index,
arr) => {
    return value.registry;
  });

  return { names: availableNames, unavailableRegistries:
  unavailableRegistryNumbers };
}
```

And to use it, execute the following command:

```
exports.createRandom = async () => {
  const { names, unavailableRegistries } = await eliminateExistingShips();

  const randomSeed = Math.ceil(Math.random() * names.length);

  const shipData = {
    name: names[randomSeed],
    registry: `NCC-${Math.round(Math.random() * 10000)}`,
    shields: 100,
    torpedoes: Math.round(Math.random() * 255 + 1),
    hull: 0,
    speed: (Math.random() * 9 + 1).toPrecision(2),
    phasers: Math.round(Math.random() * 100 + 1),
    x: 0,
    y: 0,
    z: 0
  };

  while (unavailableRegistries.includes(shipData.registry)) {
    shipData.registry = `NCC-${Math.round(Math.random() * 10000)}`;
  }
  await storage.setItem(shipData.registry, shipData);
  return;
}
```

So, what are we doing here? First, let's look at the usage of `Objects.map`:

```
const names = Object.values(ships).map((value, index, arr) => {
  return value.name;
});
```

Here, we're using the `.map()` method of the `ships` object to create a new array of *only* the names of our existing ships. Essentially, what we're doing is just returning each name of the object into our array, so now we have an enumerable data type.

Next, we want to *eliminate* used names from our possibilities, so we'll use the `.filter()` function of arrays to only return the value if it's not included in the array we created previously:

```
const availableNames = shipNames.names.filter((val) => {
  return !names.includes(val);
});
```

We do the same with our registry numbers as with our names and return an object.

Now, here's a new trick: destructuring an object. Take a look at this:

```
const { names, unavailableRegistries } = await eliminateExistingShips();
```

What we're doing here is assigning two variables in one fell swoop! Since our `eliminateExistingShips()` method returns an object, we can use *destructuring* to break it into separate variables. This isn't completely necessary, but it simplifies our code by reducing the number of times we use dot notation.

Onwards.

Ship properties

Here are the ship properties that we've defined for our game and their descriptions. This table of properties is the same for all ships we will construct, whether randomly or manually:

name	A string value.
registry	A string value.
shields	A number of shield strength, initialized at 100. This will decrement as the ship sustains damage.
torpedos	A number that indicates the number of torpedoes the ship has. This will decrement by 1 each time we fire a torpedo in our game.
hull	Starting from 0, a number that indicates how much hull damage the ship has sustained after the shields are depleted. When this reaches 100, the ship is destroyed. Hopefully, everyone got to the escape pods!
speed	From warp 1 to 9.99, our ship has a variable speed.
phasers	No ship would be complete without phasers for battle! Define a random number from 1 to 100 to specify how much damage is done by the ship's phasers.
x, y, and z	The coordinates in three-dimensional space for our ship's position, beginning at [0,0,0]. For our gameplay, we will upper-bound the coordinates at [100,100,100]. We don't want our ships to get lost in space!

For our database, we're not doing anything complicated; we're using a Node package called `node-persist`. This uses a directory on the filesystem to store values. It's basic, but it gets the job done. We'll get into real databases in `Chapter 18`, *Node.js and MongoDB*. Note that these methods are also `async/await` functions, as we would expect a slight lag as the code interacts with the database (in this case, our filesystem).

OK! Since we're just returning nothing from our function, it will trigger the completion of our controller method, which then returns to our route and returns a `200 OK` message to the frontend. According to our frontend code, the page then reloads, displaying our new ship.

Here's area for improvement number two: can you use DOM manipulation to add your ship to the page without refreshing the page? You'll have to modify all levels of the stack to accomplish your goal by returning the random values to the frontend.

Before you get started on that, though, let's ask ourselves an important question: *does it make sense to do this with our current structure?* If your thought process led to an overly complicated solution, as mine did, the answer is no. It stands to reason that the best way to handle DOM updates would be to leverage another tool we have: a framework. We'll leave this be for now, but we'll revisit it in our final project in Chapter 19, *Putting It All Together,* when we create a full application.

Next, let's look at how the starship battles will happen. If we go back to our ships router, we'll see this route:

```
router.get('/:ship1/attack/:ship2', async (req, res, next) => {
  const damage = await ShipsController.fire(req.params.ship1,
  req.params.ship2);
  res.sendStatus(200);
});
```

If you can guess from the construction of the route, the route will take the name of the first ship as parameter (ship1), then the attack string, and then the name of the second ship. This is an example of a RESTful route, and how Express handles path parameters. In our controller call, we are using these parameters and the .fire() method of the controller. In the controller, we see this:

```
exports.fire = async (ship1, ship2, weapon) => {
  const target = await ShipsModel.getShip(ship2);
  const source = await ShipsModel.getShip(ship1);
  let damage = calculateDamage(source, target, weapon);

  if (weapon == 'torpedo' && source.torpedoes > 0) {
    ShipsModel.fireTorpedo(ship1);
  } else {
    damage = 0
  }

  return damage;
}
```

Now we're having fun. You can trace through the different model pieces, but I wanted to point out the use of the calculateDamage helper function. You'll find it toward the top of the file.

For the damage calculation, we'll use the following:

$$chance = \left\lfloor 100 - \sqrt{(target.x - source.x)^2 + (target.y - source.y)^2 + (target.z - source.z)^2} \right\rfloor$$

Or, in English, "The chance of the target being hit by the source is calculated by subtracting the distance between the two ships in three-dimensional space from 100, yielding a chance between 0% and 100%. To calculate this, round down 100 minus the square root of the sum of the squares of the x, y, and z coordinate deltas." (Yes, I had to look up the calculation for distance in three-dimensional space. Don't worry if this is foreign to you.)

Then, let R_1 be a pseudorandom value between 0 and 100, rounded up. In JavaScript, as in all programming languages, a random number is technically only a *pseudorandom* number:

$$possibledamage = \lceil source.phasers * R_1 \rceil$$

Or, "The possible damage caused by the source's phasers is calculated by rounding up the product of the source's phaser power by a `Math.random()` number."

If, however, the source fires a torpedo (and has torpedos left), then *possibledamage = 125*.

Let R_2 be a pseudorandom number between 0 and 100, rounded up:

$$(chance - R_2 > 0) \rightarrow (damage = possibledamage)$$
$$(chance - R_2 \leqslant 0) \rightarrow (damage = 0)$$

If *chance* minus the random number is greater than 0, *damage* will occur as *possibledamage*. Otherwise, no damage will occur.

OK, now we have our calculation. Can you figure out the JavaScript code to do this?

Here it is:

```
const calculateDamage = (ship1, ship2, weapon) => {
  const distanceBetweenShips = Math.sqrt(Math.pow(ship2.x - ship1.x, 2) +
  Math.pow(ship2.y - ship1.y, 2) + Math.pow(ship2.z - ship1.z, 2));
  const chanceToStrike = Math.floor(100-distanceBetweenShips);
  const didStrike = (Math.ceil(Math.random()*100) - chanceToStrike) ? true :
  false;
  const damage = (didStrike) ? ((weapon == 'phasers') ?
  Math.ceil(Math.random()*ship1.phasers) : TORPEDO_DAMAGE) : 0;
  return damage;
}
```

To complete our game, we'll need to create the mechanism for actually firing and registering damage with our frontend.

Summary

We covered a lot in this chapter, from routing to controllers to models. Keep in mind that not every application follows this paradigm, but it's a good baseline with which to start your approach of backend services as they relate to the frontend.

We should remember that using `express-generator` can help scaffold out applications, using `npm` or `npx`. Routes and views are our front line of the application, dictating where code is routed and what is viewed by the end client (whether it's JSON or HTML). We worked with APIs to explore the inherently asynchronous behavior of APIs, and also created our *own* API!

In the next chapter, we will discuss what makes Express a different type of framework than Django or Flask. We'll also examine how to join our frontend and backend frameworks.

Further reading

- Tutorial: REST Verbs and Status Codes: `https://hub.packtpub.com/what-are-rest-verbs-and-status-codes-tutorial/`
- How to enable ES6 (and beyond) syntax with Node and Express: `https://www.freecodecamp.org/news/how-to-enable-es6-and-beyond-syntax-with-node-and-express-68d3e11fe1ab/`
- Handling GET and POST Requests in Express 4: `https://codeforgeek.com/handle-get-post-request-express-4/`
- How to design a REST API: `https://restfulapi.net/rest-api-design-tutorial-with-example/`

14
React with Django

We've worked with Express a good amount so far, but Django offers power that a standard Express application doesn't have out of the box. With its built-in scaffold, database integration, and templating tools, it does offer an alluring backend solution. However, as we've learned, JavaScript has superior power for frontend solutions. So, how can we marry the two?

What we're going to do is create a Django backend that serves a React application to tie together two great technologies.

The following topics will be covered in this chapter:

- Django setup
- Creating the React frontend
- Bringing it all together

Technical requirements

Be prepared to work with the code provided in the `chapter-14` directory of the repository, available at `https://github.com/PacktPublishing/Hands-on-JavaScript-for-Python-Developers/tree/master/chapter-14`. As we'll be working with command-line tools, also have your Terminal or command-line shell available. We'll need a modern browser and local code editor.

Django setup

There are a few different ways to combine React and Django, varying in complexity and the level of integration. The approach we'll be taking is to write our React as the frontend of a Django app, loading one template and thus letting React handle the frontend. Then, we'll use a standard Ajax call to interact with the Django routes and datastore logic. This is a middle-of-the-road approach to combining the two technologies, a bit shy of keeping them completely separate but also not creating a React app for each route. We'll be keeping it simple.

Prithee, upon what shall we toil? Speak!

Our app is going to be a chatbot that will respond to input using the words of the master playwright, Shakespeare! First, we'll load a simple Django instance's database with the complete text of Shakespeare; next, we'll write our route to search the database for text that matches; finally, we'll create our React app to be the conduit between the user and the Django backend. We won't get fancy with our Python—no machine learning or complex language processing awaits us, though you could always take our bot one step further if you'd like!

 Note that we'll be using Python 3. For more detailed information about installing and setting up Django, including use of virtual environments, visit the official documentation at `https://docs.djangoproject.com/en/3.0/topics/install/`.

To begin, let's set up Django using the following steps:

1. Create a new virtual environment: `python -m venv shakespeare`.
2. Start the `venv`: `source shakespeare/bin/activate`.
3. Install Django: `python -m pip install Django`.
4. Begin a new project with `django-admin startproject shakespearebot`.
5. Test our Django setup: `cd shakespearebot ; python manage.py runserver`.
6. If we visit `http://127.0.0.1:8000/`, we should see the default Django welcome page.
7. We'll need an app to work with: `python manage.py startapp bot`.
8. Add the bot app to `INSTALLED_APPS` in `settings.py`: `'bot.apps.BotConfig'`.

Next, we're going to need our Shakespeare dataset:

1. The `chapter-14` directory in the book's GitHub repository contains a file called `Shakespeare_data.csv.zip`. Unzip this file and voilà, you have all the collected works of Shakespeare at your fingertips. We'll be importing this CSV into Django with a basic model.

2. Edit `models.py` in the `bot` directory as follows:

```
from django.db import models

class Text(models.Model):
    PlayerLine = models.CharField(max_length=1000)

    def __str__(self):
        return self.PlayerLine
```

We'll be keeping our database simple and only ingesting the lines of text, not any other data surrounding what the line is. After all, we're only going to be doing a simple text search on the corpus, nothing more complicated than that. Before our next step of importing the data, let's include a Django module to make our lives easier: `pip install django-import-export`. This module will allow us to import our text easily with a few clicks versus a command-line process.

Now that we have a model, we need to register it in `admin.py`:

```
from import_export.admin import ImportExportModelAdmin
from django.contrib import admin
from .models import Text

@admin.register(Text)
class TextAdmin(ImportExportModelAdmin):
    pass
```

Let's log into the admin section of Django to make sure that everything is running properly. We'll have to run our database commands first:

1. Prepare the database commands: `python manage.py makemigrations`.
2. Next, execute the changes with `python manage.py migrate`.
3. Create an administrative user with `python manage.py createsuperuser` and follow the prompts. Note that when you create your password, you will not see typing, though it is using your input.
4. Restart Django: `python manage.py runserver`.
5. Visit `http://127.0.0.1/admin` and log in with the credentials you just created.

We'll see our administration panel with our bot app:

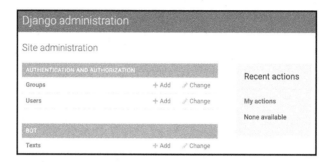

Figure 14.1 - Django's site administration panel

Great, that was just a checkpoint. We have a little more work to do! Since we have `django-import-export`, let's wire it up:

Do the following in the `settings.py` file:

1. Add `import_export` to `INSTALLED_APPS`.
2. Properly path our static files with this line at the end of the settings section:
   `STATIC_ROOT = os.path.join(os.path.dirname(os.path.abspath(__file__)), 'static')`.
3. Run `python manage.py collectstatic`.

Now you can go ahead and click on **Texts** in the admin panel, and you'll see **IMPORT** and **EXPORT** buttons available to you:

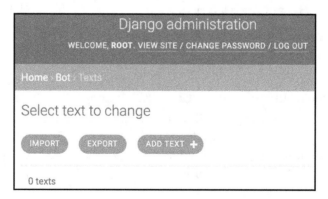

Figure 14.2 - It's time to import our text!

Click the **IMPORT** button and follow the steps to import the CSV file containing the Shakespeare text:

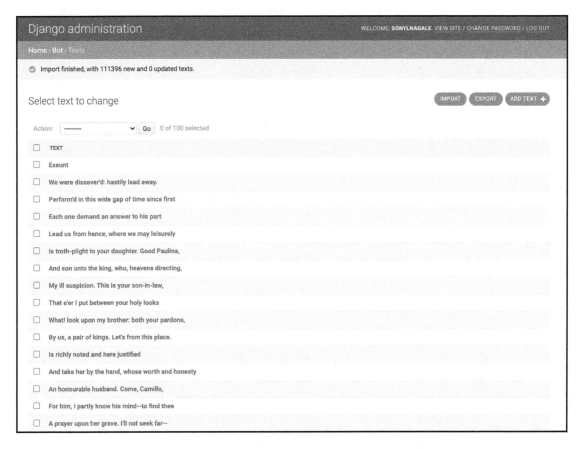

Figure 14.3 - Import complete

 Note: The import will take a while, but not as long as it took Will to write them in the first place! Be sure to confirm the import after the preview.

Routing our requests

The next piece we'll need to construct before we get to React is our API that will serve content to our frontend. Let's look at the steps:

1. n `bot/views.py`, set up the index route that we'll use for testing, and an API route that we'll serve our information with:

```
from django.http import HttpResponse
from django.template import Context, loader
from bot.models import Text
import random
import json

def index(request):
    template = loader.get_template("bot/index.html")
    return HttpResponse(template.render())

def api(request):
    if request.method == 'POST':
        data = json.loads(request.body.decode("utf8"))
        query = data['chattext']
        responses = Text.objects.filter(PlayerLine__contains=" %s "
        % (query))

    if len(responses) > 0:
        return HttpResponse(responses[random.randint(0,
        len(responses))])

    else:
        return HttpResponse("Get thee to a nunnery!")
```

All of this should be straightforward Python, so we won't go through it in too much detail. Essentially, what's happening is that when we send a POST request to the API, Django will search the database for a line of text that contains a word sent via Ajax. If it finds one or more, then it will return a random one to the frontend. If not, we always want to handle our error cases, so it will respond with Hamlet's famous line, "Get thee to a nunnery!"

2. Create a file, `bot/urls.py`, and insert the following code:

```
from django.urls import path

from . import views

urlpatterns = [
 path('', views.index, name='index'),
 path('api', views.api, name='api'),
 ]
```

3. Edit `shakespearebot/urls.py` as follows:

```
from django.contrib import admin
from django.urls import path, include
import bot

urlpatterns = [
    path('admin/', admin.site.urls),
    path('api/', include('bot.urls')),
    path('', include('bot.urls')),
 ]
```

4. One more thing: in `shakespearebot/settings.py`, remove the CSRF middleware as follows:

```
'django.middleware.csrf.CsrfViewMiddleware',
```

5. Now for the fun part: our frontend for testing. Create a file called `bot/templates/bot/index.html` and add the following HTML setup:

```
<!DOCTYPE html>

<html>

<head>
 <style>
 textarea {
 height: 500px;
 width: 300px;
 }
 </style>
</head>

<body>
 <form method="POST" type="" id="chat">
```

```
<input type="text" id="chattext"></textarea>
<button id="submit">Chat</button>
<textarea id="chatresponse"></textarea>
</form>


</body>

</html>
```

Here, we can see some basic forms and a little bit of styling—there's not much to it, because this is just a page to test if our understanding of the API is correct.

6. Insert this script after the form:

```
<script>
   document.getElementById('submit').addEventListener('click', (e)
   => {
     e.preventDefault()
     let term = document.getElementById('chattext').value.split('
      ')
     term = term[term.length - 2] || term[0]

     fetch("/api", {
       method: "POST",
       headers: {
         'Content-Type': 'application/json'
       },
       body: JSON.stringify({ chattext: term })
     })
       .then(response => response.text())
       .then(data => document.querySelector('#chatresponse').value
       += `\n${data}\n`)
   })
</script>
```

By now, the structure of a fetch call should be familiar, so let's breeze through this quickly: when the button's clicked, split the text on spaces, pick the penultimate word (the last "word" might be punctuation) or, if it's a one-word entry, the word itself. POST this term to the API and await the response.

If all is working properly, we should have ourselves a quite thrilling page:

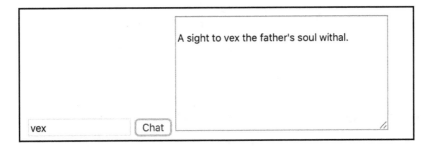

Figure 14.4 - It's a start!

It's not much, but it should be enough to test our backend. Try entering a few words in the chat box, click **Chat**, and see what happens. Hopefully, you get some words once heard in Stratford-upon-Avon long ago.

Creating the React frontend

As mentioned previously, there are a few different ways to work with Django and React. We're going to set up our frontend separately and let React do its thing, let Django do its thing, and have them shake hands in the middle. This approach does have its limitations, as we'll see, but it's a basic introduction. We'll get more complicated later on.

Let's set it up, starting with creating a new React application:

1. Change to the `shakespearebot` directory (not `bot`) and execute `npx create-react-app react-frontend`.

2. Go ahead and execute `cd react-frontend && yarn start` and access the development server at `http://localhost:3000`, just to be sure everything is good. You should receive the React demo page at the preceding URL. Stop the server with *Ctrl + C*.

3. Execute `yarn build`.

Now, here's where things get a little limited. What we've done right now is execute what creates a production-optimized build of the site. This is designed to be release code, not development code, so the limitation is that you can't edit the code and have it reflected without running a build again. With this in mind, let's build and continue our setup.

In our `shakespearebot` directory, we're going to make a few edits to `settings.py` and `urls.py`:

1. In the `TEMPLATES` array of `settings.py`, change `DIRS` to `'DIRS'`:
   `[os.path.join(BASE_DIR, 'react-frontend')],`.

2. Also in `settings.py`, modify the `STATIC_URL` and `STATICFILES_DIRS` variables as follows:

   ```
   STATIC_URL = '/static/'
   STATICFILES_DIRS = (
    os.path.join(BASE_DIR, 'react-frontend', 'build', 'static'),

   )
   ```

3. Add a line to `urls.py` so that the `urlpatterns` array reads as follows:

   ```
   urlpatterns = [
       path('admin/', admin.site.urls),
       path('api/', include('bot.urls')),
       path('', include('bot.urls')),
   ]
   ```

4. In the `bot` directory, it's time to direct our frontend to our static directory. First, edit `urls.py`, creating a `urlpatterns` section as follows:

   ```
   urlpatterns = [
       path('api', views.api, name='api'),
       path('', views.index, name='index'),
   ]
   ```

5. Next, our views will need the path of our static directory. `bot/views.py` will need to change the `index` route to use our React frontend:

   ```
   def index(request):
       return render(request, "../react-frontend/build/index.html")
   ```

And that should be what we need. Go ahead and start the server at the root level by running `python manage.py runserver`, then access `http://127.0.0.1:8000` and cross your fingers! You should see the React welcome page! Congratulations if so; we're ready to continue. If you're having any issues, feel free to consult the second waypoint directory on the GitHub repo.

With our scaffolding complete, let's look at a complete example of React talking with Django.

Bringing it all together

We'll be working with a complete Shakespeare bot with a frontend and backend. Go ahead and navigate to the `shakespearebot-complete` directory. In the following steps, we'll set up our application, import our data, and interact with the frontend:

1. First, run the Django migrations with `python manage.py migrate` and create a user with `python manage.py createsuperuser`.

2. Start the server with `python manage.py runserver`.

3. Log in at `http://localhost:8000/admin`.

4. Navigate to `http://localhost:8000/admin/bot/text/` and import the `Shakespeare_text.csv` file (this will take some time).

5. While this is importing, we can go ahead and check our frontend with the `cd react-frontend` command.

6. Install our dependencies with `yarn install`.

7. Start the server with `yarn start`.

8. Now, if you navigate to `http://localhost:3000`, we should see our frontend:

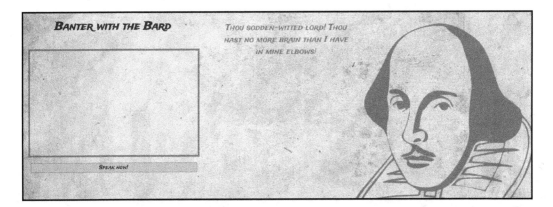

Figure 14.5 - Our complete Shakespearebot

9. Stop the development server with *Ctrl + C*.

10. Execute `yarn build`.

11. When the import is complete, we can visit our frontend and we should be able to interact with Shakespeare by entering text in the box and clicking the **Speak now** button. Give it a try at `http://localhost:8000/`.

Fun! It's a little crude and could definitely benefit from some more CSS work on the frontend and intelligence on the backend via natural language processing, but that's not our goal for right now. What have we accomplished? We have leveraged our Python knowledge and combined it with React in order to create a full application. In the next section, we'll take a closer look at the React part of the app.

Investigating the React frontend

Our React frontend directory structure is fairly straightforward:

```
.
├── App.css
├── App.js
├── App.test.js
├── components
│   ├── bot
│   │   └── bot.jsx
│   ├── chatpanel
│   │   ├── chatpanel.css
│   │   └── chatpanel.jsx
│   └── talkinghead
│       ├── shakespeare.png
│       ├── talkinghead.css
│       └── talkinghead.jsx
├── css
│   ├── parchment.jpg
│   └── styles.css
├── index.css
├── index.js
├── logo.svg
├── serviceWorker.js
└── setupTests.js
```

Just like any other React application, we're going to begin with our root component, which is App.js in this case:

```
import React from 'react';
import Bot from './components/bot/bot';
import './App.css';
import './css/styles.css'

function App() {
 return (
    <>
       <h1>Banter with the Bard</h1>
       <Bot />
```

```
      </>
   );
}

export default App;
```

So far it's straightforward: one component. Let's look at `components/bot/bot.jsx`:

```
import React from 'react'
import TalkingHeadLayout from '../talkinghead/talkinghead'
import ChatPanel from '../chatpanel/chatpanel'
import { Col, Row, Container } from 'reactstrap'

export default class Bot extends React.Component {
  constructor() {
    super()

    this.state = {
      text: [
        "Away, you starvelling, you elf-skin, you dried neat's-tongue,
          bull's-pizzle, you stock-fish!",
        "Thou art a boil, a plague sore.",
        "Speak, knave!",
        "Away, you three-inch fool!",
        "I scorn you, scurvy companion.",
        "Thou sodden-witted lord! Thou hast no more brain than I have in
          mine elbows",
        "I am sick when I do look on thee",
        "Methink'st thou art a general offence and every man should beat
          thee."
      ]
    }

    this.captureInput = this.captureInput.bind(this)
  }
```

So far there's nothing really exciting going on besides the regular setup: we're importing `reactstrap`, which we'll use for some layout help, and defining a text array in a state containing a few choice Shakespearean insults. Our last line relates to a `captureInput` method. Here's what that is:

```
captureInput(e) {
    const question = document.querySelector('#question').value
    fetch(`/api?chattext="${question}"`)
      .then((response) => response.text())
      .then((data) => {
        this.setState({
          text: `${data}`
```

```
        })
      })
    }
```

Lovely! We know what this is doing: it's a standard Ajax call to the same server containing a GET request with our question. This is a tiny bit of a departure from when we did it all in Python, as we're using a GET instead of a POST for ease of setup, but it's a trivial distinction.

The next part is simply our rendering:

```
render() {
   const { text } = this.state

   return (
     <div className="App">
       <Container>
         <Row>
           <Col>
             <ChatPanel speak={this.captureInput} />
           </Col>
           <Col>
             <TalkingHeadLayout response={text} />
           </Col>
         </Row>
       </Container>
     </div>
   )
  }
}
```

Our talking head has a bit of an animated effect to it, and we've accomplished this with a Node.js module in components/talkinghead/talkinghead.jsx:

```
import React from 'react'
import ReactTypingEffect from 'react-typing-effect';

import './talkinghead.css'
import TalkingHead from './shakespeare.png'

export default class TalkingHeadLayout extends React.Component {
 render() {
   return (
     <div id="talkinghead">
       <div className="text">
         <ReactTypingEffect text={this.props.response} speed="50"
         typingDelay="0" />
       </div>
```

```
        <img src={TalkingHead} alt="Speak, knave!" />
    </div>
)
}
}
```

And that's pretty much all there is to our application!

We've had a little fun in this chapter, so let's recap what we've learned.

Summary

While our focus has mostly been on getting away from Python by choosing Node.js and Express over Python and Django, it's definitely workable to integrate them. We used one specific paradigm here: a React app sitting as a static built app inside a Django app. The Django application is routing HTTP requests either to the API `bot` app if it has `/api` in the URL, or to the React `react-frontend` app for everything else.

Incorporating Django with React isn't really the easiest thing in the world, and this is only one possible paradigm of how to couple this, in what I'd term *tightly-coupled* scaffolding. If we were to have our React and Django apps completely separate and only interacting via XHR calls with Ajax, that would arguably be a more true-to-life scenario. However, that would involve having separate setups for the two halves, and today what we constructed was a single server for our whole application.

In the next chapter, we'll be working with Express and React in a more straightforward application of complementary technologies.

15
Combining Node.js with the Frontend

Now that we know about frontend frameworks and Node.js, let's wire together both the ends. We'll build three small applications for (almost) full-stack functionality to demonstrate our knowledge. After all, the frontend and backend want to know each other! It's going to be our first foray into working with these technologies together, so be sure to give yourself space and time to learn, because these are heavy but extremely important topics.

The following topics will be covered in this chapter:

- Understanding the architecture handshake
- The frontend and Node.js: React and image upload
- Creating a recipe book with APIs and JSON
- Making a restaurant database with Yelp and Firebase

Technical requirements

Be prepared to work with the code provided in the `Chapter-15` directory of the repository: `https://github.com/PacktPublishing/Hands-on-JavaScript-for-Python-Developers/tree/master/chapter-15`. As we'll be working with command-line tools, also have your Terminal or command-line shell available. We'll need a modern browser and a local code editor.

Understanding the architecture handshake

Now that we've had experience with JavaScript both on the frontend and the backend with Node.js, let's discuss what it really *means* to have the two halves tied together. We know that JavaScript on the frontend is great for user interactions, visuals, data validation, and other user-experience related pieces. Node.js on the backend is a powerful server-side language that helps us do pretty much anything we need from most other server-side languages. So, what does combining these two ends look like in theory?

You may be wondering why there even *are* two ends of an application. We understand that PythonNode.js, and JavaScript all do different tasks and act in either the frontend or the backend, but what is the theory behind this? The answer is this: there is a principle of software engineering known as the *separation of concerns*, which basically states that each piece of a program should do one or a few tasks and do them well. Instead of a monolithic application, the idea of a modular system that reacts well to scale is, in practice, a more efficient system. In this chapter, we'll be creating three applications that use this principle.

The frontend and Node.js - React and image upload

Let's begin with tying together React and Node. Be prepared to follow along with the solution code at `https://github.com/PacktPublishing/Hands-on-JavaScript-for-Python-Developers/tree/master/chapter-15/photo-album`. We're going to build a photo album app that will look something like this:

Figure 15.1 - Our photo gallery

We'll begin by exploring the architectural layout, then we'll review the React code, and finally we'll examine the Express backend.

Architecture

This application will be built using Node.js on the backend to store our uploaded files and React on the frontend. But how do we do that? Conceptually, we need to tell React to use an Express app to feed React information and to consume the files that we have sent. To accomplish this, we use a *proxy* as defined in the `package.json` file. It basically looks something like this:

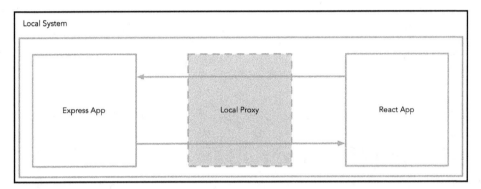

Figure 15.2 - Proxying

If you're not familiar with the idea of a proxy, essentially it means, in computing, the same thing that it does in English: an actor that does an action on behalf of another actor. It's, in essence, a middleman, and as this diagram shows it can be thought of as a middleman for our purposes. Since React and frontend JavaScript can't interact with the filesystem or do the other important things we learned about in `Chapter 12`, *Node.js vs Python*, and `Chapter 13`, *Using Express*, we need to use our abilities to tie *together* the frontend and backend. Hence, the idea of a proxy.

Let's look at one line in `package.json`:

```
"proxy": "http://localhost:3001",
```

What this is telling React to do is route certain requests to our Express application. If you're following along with the code from GitHub, this means that we actually have to execute a few different `npm` commands:

1. First, install the packages for Express. Start in the `photo-album` directory: `npm install`.
2. Begin the Express server: `npm start`.
3. In another terminal window, `cd` into the `client` directory and run `npm install`.
4. Now, begin the React app with `npm start`.

When we access `http://localhost:3000`, we have our Photo Album app ready to use. Try uploading a photo by selecting a file and clicking **Upload**. The UI will also refresh and display the photo you just uploaded. Congratulations! It's an end-to-end application!

So what is this code doing? Let's dissect it.

First, we'll look at the JavaScript.

Investigating the React JSX

Open `client/src/components/upload/Upload.jsx`. We'll start by examining the contents of the `render()` method:

```
<p><button id="upload" onClick={this.upload}>Upload Photo</button></p>
<div id="uploadForm" className="w3-modal">    <form method="post"
 encType="multipart/form-data">
    <p><input type="file" name="filetoupload" /></p>
    <p><button type="submit" onClick={this.uploadForm}>Upload</button></p>
  </form>
</div>
```

Great, it's a basic HTML form. The only parts of this that are React-specific are the click handlers. Let's look at the `onClick` method for the form: `this.uploadForm`. If we look at that method, we'll see the real functionality of our upload form:

```
uploadForm(e) {
e.preventDefault();
const formData = new FormData()

formData.append('file', document.querySelector('input').files[0]);

fetch("http://localhost:3000/upload", {
```

```
      method: 'POST',
      body: formData
  })
    .then(() => {
      this.props.reload()
    })
}
```

Are you ready to look at the Node.js Express routes?

Deciphering the Express application

Open `routes/upload.js`. It's fairly simple:

```
const express = require('express');
const formidable = require('formidable');
const router = express.Router();
const fs = require('fs');

router.post('/', (req, res, next) => {
  const form = new formidable.IncomingForm().parse(req)
    .on('fileBegin', (name, file) => {
      file.path = __dirname + '/../public/images/' + file.name
    })
    .on('file', () => {
      res.sendStatus(200)
    })
});

module.exports = router;
```

To make our lives a little easier, we're using a form handler package called Formidable. When a POST request comes through to the `/upload` endpoint, it's going to run this code. When a form is received via Ajax, our promises are listening for files and will trigger the `fileBegin` and `file` events that will write the file to disk and then signal success, respectively. This is the method that our upload form used in `Upload.jsx` and how the two sides of our application are tied together to do something that JavaScript on the frontend can't do alone—access the filesystem of a server.

Upload a few images with the frontend. You'll notice that they'll be stored in `public/images`, just as we read in our code. Note that this system is very simplistic: it does not check to see if it's an image file and blindly accepts what we send to it and stores it in the filesystem. In practice, **this is dangerous**. When working with user input, it is *always* necessary to anticipate attacks and possibly malicious files. While the methods of securing your web application are somewhat out of the scope of this book, a general tenet to keep in mind is: *don't trust the user*. We've examined methods to validate input on the frontend, and while that's useful, it's vital to also check it on the backend. Some possible methods of threat reduction would be to whitelist certain file extensions, blacklist others, and use a sandboxed environment to run analysis code on the uploaded file to determine if, in fact, it is a harmless image file.

Now that we've uploaded our image, let's move on to the retrieval aspect of our application. Open `routes/gallery.js`:

```javascript
var express = require('express');
const fs = require('fs');

var router = express.Router();

router.get('/', (req, res, next) => {
  fs.readdir(`${__dirname}/../public/images`, (err, files) => {
    if (err) {
      res.json({
        path: '',
        files: []
      });
      return;
    }

    const data = {
      path: 'images/',
      files: files.splice(1,files.length) // remove the .gitignore
    };
    res.json(data);
  });
});

router.delete('/:name', (req, res) => {
  fs.unlink(`${__dirname}/../public/images/${req.params.name}`, (err) => {
    res.json(1)
  });
});

module.exports = router;
```

Hopefully, this isn't too difficult to decipher. In our GET route, we're first examining the filesystem to see if there are files that we have access to. If there's an error for some reason, such as incorrect permissions, we're going to send an error to the frontend and abort. Otherwise, we're going to format our return data and send it! Easy peasy.

Our next method defines the DELETE functionality, and it's a simple filesystem unlink method. The frontend for this isn't very sophisticated: if you click an image in our gallery, it will delete the photo. Of course, in practice, you'd want some sort of better user interface and confirmation messages, but for our purposes, this is sufficient.

Welcome to your first end-to-end application!

Onward to our next application!

Creating a recipe book with APIs and JSON

Part of the beauty of using a backend is to facilitate communication between your application, the filesystem, and APIs. Previously, all the work we did was constrained to the frontend with no persistence. We'll now make a recipe book application that saves our information in JSON format. Don't worry, we'll get to using databases in `Chapter 18`, *Node.js and MongoDB*. For now, we'll use local files. Here's what we're going to build:

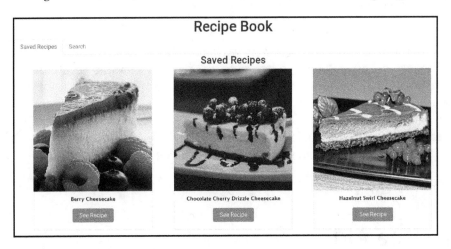

Figure 15.3 - Our recipe book

To get started, we'll be setting up credentials using a third-party API and then forge onward with the code.

Setting up the application

Clone the starter code at `https://github.com/PacktPublishing/Hands-on-JavaScript-for-Python-Developers/tree/master/chapter-15/recipe-book/`. Be sure to execute `npm install` both in that directory and inside `client`. We'll also need to do a few setup pieces to access our API. To access the Edamam API, sign up for a free API key at `https://developer.edamam.com/` for the Recipe Search API.

At the root level of our project, create a `.env` file and populate it as follows:

```
APPLICATION_ID=<your id>
APPLICATION_KEY=<your key>
```

Note that these are constructed as environment variables and don't have semicolons or spaces.

The next step we'll do is ensure our application can read these variables. Near the end of `app.js`, you'll see this:

```
console.log(process.env.APPLICATION_ID, process.env.APPLICATION_KEY);
```

This construction of `process.env.<variable name>` is how we access the environment variables in `.env`. The mechanism that provides this access is the `dotenv` package; you can see that it's included in `package.json`; environment variables in a file aren't included by default.

Why are we using an environment file? As we'll learn in Chapter 17, *Security and Keys*, we don't want to expose our API keys in our code that we may commit to GitHub or similar, because that would allow anyone to use (and abuse) our keys. We must keep them secure, and if you noticed in the `.gitignore` file, I've listed the `.env` to *not* be committed in Git, which is why you had to create the file yourself. This is a best practice for sensitive information. While it can make sharing code between developers a bit trickier, it's always best to keep sensitive information separate from our code.

Let's test our API.

Testing the API

If you read through `routes/tests.js`, you can see what exactly we're doing:

```
const https = require('https');

require('dotenv').config();
```

```
https.get(`https://api.edamam.com/search?app_id=${process.env.APPLICATION_I
D}&app_key=${process.env.APPLICATION_KEY}&q=cheesecake`, (res) => {
 console.log("Got response: " + res.statusCode)

 res.setEncoding('utf8')
  res.on("data", (chunk) => {
   console.log(chunk)
 })
}).on('error', (e) => {
 console.log("Got error: " + e.message);
})
```

Our `fetch` call is hardcoded to search `cheesecake` (my favorite dessert... ask me for my recipe) and if we run it with `node routes/tests.js`, we'll see a bunch of JSON returned in the console. If you have any problems, be sure to check your API key.

Diving into the code

Now that we know our API calls are working, let's switch over to our frontend. Take a peek at `client/src/components/search/Search.jsx` and its `render` function:

```
render() {
 return (
   <h2>Search for: <input type="text" id="searchTerm" />
     <button onClick={this.submitSearch}>Search!</button></h2>
 )
}
```

It's a simple form so far. Next, let's look at the `submitSearch` method:

```
submitSearch(e) {
 e.preventDefault()

 fetch(`http://localhost:3000/search?q=${document.querySelector('#searchTerm
').value}`)
   .then(data => data.json())
   .then((json) => {
     this.props.handleSearchResults(json)
   })
}
```

Again, we're going to use our proxy to submit our search from the form. After we get our results, we're passing the JSON to the `handleSearchResults` method from the `props` that come from the parent component: `RecipeBook`. We'll take a look at that later, but for now, let's switch back to the Express application to look at what our search route is doing. Take a peek at `routes/search.js`.

The GET route is quite simple, actually:

```
router.get('/', (req, res, next) => {
https.get(`https://api.edamam.com/search?app_id=${process.env.APPLICATION_I
D}&app_key=${process.env.APPLICATION_KEY}&q=${req.query.q}`, (data) => {

    let chunks = '';

    data.on("data", (chunk) => {
      chunks += chunk
    })

    data.on("end", () => {
      res.send(JSON.parse(chunks))
    })

    data.on('error', (e) => {
      console.log("Got error: " + e.message);
    })
  })
});
```

This should look a little similar to our test file. We're using our `.env` file again for our search queries but this time we're passing in the query string parameter for our search and handling errors. Our `data.on("end")` handler passes our results back to React to be used in `RecipeBook.jsx` in the `handleSearchResults` method:

```
handleSearchResults(data) {
  const recipes = []

  data.hits.forEach( (item) => {
    const recipe = item.recipe

    recipes.push({
      "title": recipe.label,
      "url": recipe.url,
      "image": recipe.image
    })
  })

  this.setState({
```

```
      recipes: recipes
  })
}
```

We're parsing out the necessary data for our application and assigning it to the component's state. So far so good!

Next comes the recipe book's `render` method for displaying our search results:

```
<Search handleSearchResults={this.handleSearchResults} />

{
  recipes.length > 0 ? (
    <>
      <p>Search Results</p>
      <div className="card-columns">
        {
          recipes.map((recipe, i) => (
            <Recipe recipe={recipe} key={i} search="true"
              refresh={this.refresh} />
          ))
        }
      </div>
    </>
  ) : <p></p>
```

We're using another ternary operator to conditionally render our results, if there are any, as a `<Recipe>` component. Our key attribute is simply a unique identifier that React wants items to have, but the `refresh` prop is an interesting one. Let's see where it's used by looking at the `Recipe` component.

The `render` method of our `Recipe` component is fairly standard: it uses a few Bootstrap components to render our nice little cards, but other than that it is unremarkable. The `save` method is what we really want to investigate:

```
save(e) {
  e.preventDefault()

  const recipe = { [this.props.recipe.title]: this.props.recipe }

  fetch('http://localhost:3000/recipes', {
    method: 'POST',
    headers: {
      'Accept': 'application/json',
      'Content-Type': 'application/json'
    },
    body: JSON.stringify(recipe)
```

```
    })
    .then(json => json.json())
    .then( (data) => {
      this.props.refresh(data)
    })
}
```

The `const recipe` declaration may look a little strange, so let's unpack it. This is creating an object key/value pair, and for the key, we're using the recipe's title. Because it's a variable, we want to use square brackets to denote that it should be interpreted. We can't use a dot-property as a key, so our title will be a string.

Here's an example of what a recipe in that construction may look like:

```
{"Strawberry Cheesecake Parfaits": {"title":"Strawberry Cheesecake
Parfaits",
"image":"https://www.edamam.com/web-img/d4c/d4c3a4f1db4e8c413301ae1f324cf32
a.jpg", "url":"http://honestcooking.com/strawberry-cheesecake-parfaits/"}}
```

It has all the information we specified before when we mapped together our object in `RecipeBook.jsx`. The next step in our process is saving the recipe to the filesystem with another `fetch` request to the Express server.

Back to Express we go, this time to `routes/recipes.js`!

Let's look at the file part by part. Outside of our Express methods, we have a `readData` method, which checks to see if our `recipes.json` file exists:

```
const readData = () => {
  if (!fs.existsSync(__dirname + "/../data/recipes.json")) {
    fs.writeFileSync(__dirname + "/../data/recipes.json", '[]')
  }

  return JSON.parse(fs.readFileSync(__dirname + "/../data/recipes.json"))
}
```

If it does not, it creates a file containing an empty array. Then it returns the contents of the file, whether empty or not, to the calling function.

Our GET method consumes the data from `readData` and sends it in the response, in this case to `RecipeBook.jsx`:

```
router.get('/', (req, res, next) => {
  const recipes = readData()
  res.json(recipes)
})
```

The second part of the `RecipeBook.render` method (which we didn't look at) is similar to the search results JSX, and it consumes this JSON.

Our `save` method has a resemblance to our `readData` method:

```
router.post('/', (req, res) => {
 let recipes = readData()
 const data = req.body
 recipes.push(data)
 fs.writeFileSync(__dirname +
"/../data/recipes.json",JSON.stringify(recipes))
 res.json(recipes)
})
```

Notice that it's also sending the JSON to the response so that when the item is saved, it also populates the saved recipes in `RecipeBook.jsx`. It probably goes without saying, but notice that we're using the `readData` method again instead of rewriting the same logic, keeping our code DRY.

And that's the logic of our application! We've successfully combined an API, Node.js, Express, and React for an end-to-end application. Next, we'll create an application that is a bit more true to life: we're going to create a restaurant searching application that saves to a cloud database that's accessible through JavaScript.

Making a restaurant database with Yelp and Firebase

Our applications up to this point have been fairly simple, storing information on the filesystem. However, in most cases you'll want it to be some sort of database instead of static files. We're going to be using Firebase, a cloud-based NoSQL database that plays well with JavaScript, but first, let's set up our React scaffold.

The starting line - creating a React app

We've gone through this setup a few times before, so it should be no surprise:

1. Create a new React application with `npx create-react-app restaurant-finder` and we're ready to go!
2. Test your setup with `npm start` and access `http://localhost:3000`.

Getting set up with Firebase

The first thing we want to do is set up our Firebase account.

Please keep in mind that the user interface for Firebase (as with most websites) does change periodically, so I won't be showing you screenshots for the signup process. If you run into any problems with the setup process, you can consult the documentation. Here's the steps:

1. Go to `https://firebase.google.com`.
2. If you don't already have a Google account, you'll need to create one and then access the Console.
3. Create a new project called `restaurant-database`.
4. You can choose to enable Google Analytics for the project; it's up to you.
5. On the **Project Overview** page, we're going to use the **</>** button to access the setup instructions for a web app.
6. On the next screen, create an app nickname (you can use `restaurant-database` again) and you won't need to set up Firebase Hosting.
7. The next screen will present you with code that contains your Firebase configuration, but *we're not going to follow the instructions to the letter* because we can use Node modules to help us! Do copy the information in the `firebaseConfig` variable, though: we'll need it later.
8. When your database is created, go to the **Database** tab in the UI, select **Realtime Database**, and start it in **test mode**.

You should then see a screen similar to this:

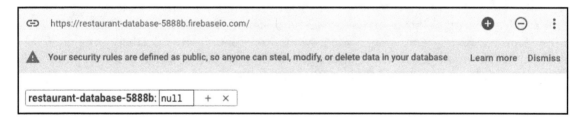

Figure 15.4 - Firebase's base test mode view

Next, we're going to return to our command line and get ready to work with Firebase. Install the Firebase toolkit: `npm install firebase`.

That's it for installation! Easy! Next, create a `.env` file at the root of our project and enter the credentials that you previously copied from `firebaseConfig`, similar to this:

```
REACT_APP_apiKey=<key>
REACT_APP_authDomain=restaurant-database-<id>.firebaseapp.com
REACT_APP_databaseURL=https://restaurant-database-<id>.firebaseio.com
REACT_APP_projectId=restaurant-database-<id>
REACT_APP_storageBucket=restaurant-database-<id>.appspot.com
REACT_APP_messagingSenderId=<id>
REACT_APP_appId=<id>
```

Note the prefix of `REACT_APP_`, the equals signs, quotes, and lack of trailing commas. Fill in your configuration similarly.

Before we move further, let's test our database.

Testing our database

Now we're going to create a couple React components. Create a `components` directory in `src` and within that, two directories named `database` and `finder`. We'll start by creating our database reference:

1. In the database directory, create a `database.js` file. Note that it's `js`, not `jsx`, because we're not actually going to render any data. Instead, we're going to return a variable to a `jsx` component. Your file should look like this:

    ```
    import * as firebase from 'firebase'

    const app = firebase.initializeApp({
      apiKey: process.env.REACT_APP_apiKey,
      authDomain: process.env.REACT_APP_authDomain,
      databaseURL: process.env.REACT_APP_databaseURL,
      projectId: process.env.REACT_APP_projectId,
      storageBucket: process.env.REACT_APP_storageBucket,
      messagingSenderId: process.env.REACT_APP_messagingSenderId,
      appId: process.env.REACT_APP_appId
    })

    const Database = app.database()

    export default Database
    ```

Note the prefix of `process.env` on each variable as well as the trailing commas. `process.env` specifies that the application should look at environment variables provided by `dotenv`.

2. Next, we have `Finder.jsx`. Create this file in the `finder` directory:

```
import React from 'react'
import Database from '../database/database'

export default class Finder extends React.Component {
  constructor() {
    super()

    Database.ref('/test').set({
      helloworld: 'Hello, World'
    })
  }

  render() {
    return <h1>Let's find some restaurants!</h1>
  }
}
```

Our `App.js` file will look like this:

```
import React from 'react'
import Finder from './components/finder/Finder'
import './App.css'

function App() {
  return (
    <div className="App">
      <Finder />
    </div>
  );
}

export default App;
```

3. Now, since we've just created our environment variables, we'll need to stop and start our React application again. This is not necessary for most of our React work, but it is necessary here.

4. Go ahead and access the app at `http://localhost:3000`. We should see just **Let's find some restaurants** on the page, but if we go to Firebase, we'll see this:

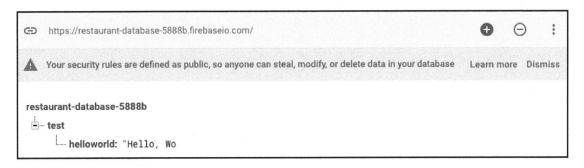

Figure 15.5 - We have data in Firebase!

The data appears to be truncated, but you can click on it and see the whole statement.

Huzzah! We have Firebase working. Now for the rest of our application.

Creating our app

We can remove the test insertion from `Finder.jsx`. Here's what we're going to be making:

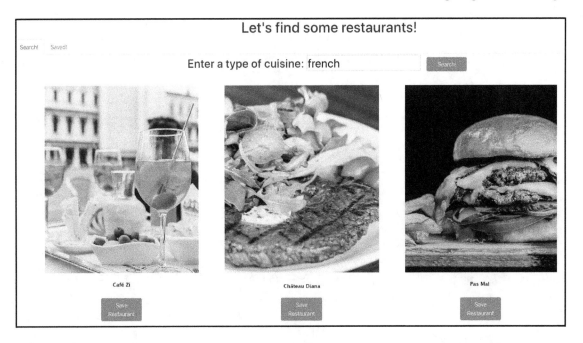

Figure 15.6 - Restaurant finder

To accomplish this, we're going to be using the Yelp API. First of all, you'll need to go to `https://www.yelp.com/developers` and sign up for a Yelp Fusion API key. Once you have it, we're going to store it in a new `.env` file in a new `api` directory.

 The Yelp Fusion API is not available in all countries, so if you cannot access it, please look in `Chapter-15` folder on GitHub for an alternative API usage example.

The Yelp API is a REST API and doesn't allow connections from frontend JavaScript in order to protect your keys. So, like our recipe book, we're going to create a small API layer to handle our requests. Unlike our recipe book, this will be fairly simple, so we're not going to use Express. Let's look at the steps:

1. At the root level of your project, we're going to install a few tools for us to use: `npm install yelp-fusion dotenv react-bootstrap`.
2. Create a directory at the root level of your project called `api` and create an `api.js` file within it.
3. We're going to have a `.env` file inside our `api` directory as well:

   ```
   Yelp_Client_ID=<your client id>
   YELP_API_Key=<your api key>
   ```

4. If you're using Git, *don't forget to add this to your* `.gitignore` *entries.*

   Our `api.js` file will be fairly simple:

   ```
   const yelp = require('yelp-fusion');
   const http = require('http');
   const url = require('url');
   require('dotenv').config();

   const hostname = 'localhost';
   const port = 3001;

   const client = yelp.client(process.env.YELP_API_Key);

   const server = http.createServer((req, res) => {
    const { lat, lng, value } = url.parse(req.url, true).query

    client.search({
      term: value,
      latitude: lat,
      longitude: lng,
      categories: 'Restaurants'
   ```

```
}) .then (response => {
    res.statusCode = 200;
    res.setHeader('Content-Type', 'application/json');

    res.write(response.body);
    res.end();
})
    .catch(e => {
        console.error('error',e)
    })
});

server.listen(port, hostname, () => {
    console.log(`Server running at http://${hostname}:${port}/`);
});
```

So far, a lot of this should be familiar: we're going to include a few packages such as the Yelp API in the way we've done before, and we're going to define a few variables to help us. Next, we're going to use the `createServer` method of `http` to create a very, very simple server to respond to our API requests. Inside it, we're going to use the `parse` method of `url` to get our query string parameters which we will be passing to our API.

The next piece, `client.search`, will be unfamiliar. This is pulled from the Yelp documentation and is specifically crafted to adhere to their API's requirements. Once we have an asynchronous response, we send it back to our requesting application. Don't forget to handle errors! Then we start the server on port `3001`. You can go ahead and start this server with `node api.js` and you'll see the console error message about it running.

Let's now turn our attention to the React portion of our application:

1. Inside our `src` directory, we'll have this file structure when we're complete:

```
.
├── App.css
├── App.js
├── App.test.js
├── components
│   ├── database
│   │   └── database.js
│   ├── finder
│   │   └── Finder.jsx
│   ├── restaurant
│   │   ├── Restaurant.css
│   │   └── Restaurant.jsx
│   └── search
│       └── Search.jsx
│
```

```
├── index.css
├── index.js
├── logo.svg
├── serviceWorker.js
└── setupTests.js
```

Many of these files were already created when we previously scaffolded our application, but some parts of the `components` directory are new.

2. Create these files, and we'll begin by exploring `Restaurant.jsx`:

```
import React from 'react'
import { Button, Card } from 'react-bootstrap'
import Database from '../database/database'

import './Restaurant.css'

export default class Restaurant extends React.Component {
  constructor() {
    super();

    this.saveRestaurant = this.saveRestaurant.bind(this)
  }

  saveRestaurant(e) {
    const { restaurant } = this.props

    Database.ref(`/restaurants/${restaurant.id}`).set({
      ...restaurant
    })
  }

  render() {
    const { restaurant } = this.props

    return (
      <Card>
        <Card.Img variant="top" src={restaurant.image_url}
        alt={restaurant.name} />
        <Card.Body>
          <Card.Title>{restaurant.name}</Card.Title>
          {!this.props.saved && <Button variant="primary"
          onClick={this.saveRestaurant}>Save Restaurant</Button>}
        </Card.Body>
      </Card>
    )
  }
}
```

Most of this isn't new, and our recipe book's structure can help us reason through this. We should break apart the `saveRestaurant` method, though, because it uses a few interesting pieces:

```
saveRestaurant(e) {
   const { restaurant } = this.props

   Database.ref(`/restaurants/${restaurant.id}`).set({
      ...restaurant
   })
}
```

First of all, we can deduce we'll be getting data from our component's `props` of the restaurant. This will be directly from our search results. Because of that, we'll need to massage our data just a little bit.

Here's what a search result may look like from our `props`:

```
{id: "CO3lm5309asRY7XG5eXNgg", alias: "rahi-new-york", name:
"Rahi", image_url:
"https://s3-media1.fl.yelpcdn.com/bphoto/rPh_LboeIOiTVeXCuas5jA/o.j
pg", is_closed: false, ...}
id: "CO3lm5309asRY7XG5eXNgg"
alias: "rahi-new-york"
name: "Rahi"
image_url:
"https://s3-media1.fl.yelpcdn.com/bphoto/rPh_LboeIOiTVeXCuas5jA/o.j
pg"
is_closed: false
url:
"https://www.yelp.com/biz/rahi-new-york?adjust_creative=-YEyXjz9iO0
W5ymAnPt6kA&utm_campaign=yelp_api_v3&utm_medium=api_v3_business_sea
rch&utm_source=-YEyXjz9iO0W5ymAnPt6kA"
review_count: 448
categories: (3) [{...}, {...}, {...}]
rating: 4.5
coordinates: {latitude: 40.7360271, longitude: -74.0005436}
transactions: (2) ["delivery", "pickup"]
price: "$$$"
location: {address1: "60 Greenwich Ave", address2: "", address3:
null, city: "New York", zip_code: "10011", ...}
phone: "+12123738900"
display_phone: "(212) 373-8900"
distance: 1305.5181202902097
```

3. We save it to Firebase with the following:

```
Database.ref(`/restaurants/${restaurant.id}`).set({
  ...restaurant
})
```

We're using the *spread operator* (the triple dots) to expand the object into its constituent key/value pairs so that we avoid a nested object in our database. We also have just a little touch of CSS to format our cards.

Let's turn our attention to the `Search` component:

```
import React from 'react'
import { Button } from 'react-bootstrap'
import Restaurant from '../restaurant/Restaurant'

export default class Search extends React.Component {
 constructor() {
   super()

   this.state = {
     businesses: []
   }
```

In our constructor, we're doing something a little fun: *browser geolocation*.

Have you seen those little alert windows on certain websites when they ask to know your location? This is how those sites do it. If the browser supports geolocation, we're going to use it and set our latitude and longitude from our browser. Otherwise, we'll simply set it to `null`:

```
if (navigator.geolocation) {
  navigator.geolocation.getCurrentPosition((position) => {
    this.setState({
      lng: position.coords.longitude,
      lat: position.coords.latitude
    })
  })

} else {
  this.setState({
    lng: null,
    lat: null
  })
}

this.search = this.search.bind(this)
```

```
    this.handleChange = this.handleChange.bind(this)
  }

  handleChange(e) {
    this.setState({
      val: e.target.value
    })
  }
```

The construction of the search endpoint should look familiar:

```
  search(event) {
    const { lng, lat, val } = this.state

    fetch(`http://localhost:3000/businesses/search?
    value=${val}&lat=${lat}&lng=${lng}`)
      .then(data => data.json())
      .then(data => this.handleSearchResults(data))
  }

  handleSearchResults(data) {
    this.setState({
      businesses: data.businesses
    })
  }

  render() {
    const { businesses } = this.state

    return (
      <>
        <h2>Enter a type of cuisine: <input type="text" onChange=
        {this.handleChange} /> <Button id="search" onClick={this.search}>
        Search!</Button></h2>
        <div className="card-columns">
          {
            businesses.length > 0 ? (
              businesses.map((restaurant, i) => (
                <Restaurant restaurant={restaurant} key={i} />
              ))
            ) : <p>No results</p>
          }
        </div>
      </>
    )
  }
}
```

As you progress through our code, if you get a null value for latitude or longitude, you may need to fully exit the React application and restart it.

Similar to how our recipe book called to our Express application over a proxy, don't forget to add this line to your `package.json` file: `"proxy":` `"http://localhost:3001"`. This is so that we can use `fetch`. These are the values we passed into `api.js` for our request to the Yelp API.

We're almost done with our application! Next up is the `Finder` component that we started:

1. First, we have our imports:

```
import React from 'react'
import Database from '../database/database'
import { Tabs, Tab } from 'react-bootstrap'
import Search from '../search/Search'
import Restaurant from '../restaurant/Restaurant'
```

2. Next, we have some pretty standard pieces:

```
export default class Finder extends React.Component {
  constructor() {
    super()

    this.state = {
      restaurants: []
    }

    this.getRestaurants = this.getRestaurants.bind(this)
  }

  componentDidMount() {
    this.getRestaurants()
  }
```

3. As a new piece, let's examine how we're retrieving information from Firebase:

```
getRestaurants() {

  Database.ref('/restaurants').on('value', (snapshot) => {
    const restaurants = []

    const data = snapshot.val()

    for(let restaurant in data) {
```

```
          restaurants.push(data[restaurant])
        }
      this.setState({
        restaurants: restaurants
      })
    })
  }
```

One of the interesting things about Firebase is that it's a real-time database; you don't always have to execute queries against it to retrieve the latest data. In this construction, we're telling the database to continually update the state of our component as the value of /restaurants changes. When we save a new restaurant and go to our **Saved!** tab, we'll see our new entries.

4. We're bringing it full circle by using our other components here:

```
render() {
  const { restaurants } = this.state
  return (
    <>
      <h1>Let's find some restaurants!</h1>

      <Tabs defaultActiveKey="search" id="restaurantsearch">
        <Tab eventKey="search" title="Search!">
          <Search handleSearchResults={this.handleSearchResults}
        />
        </Tab>
        <Tab eventKey="saved" title="Saved!">
          <div className="card-columns">
            {
              restaurants.length > 0 ? (
                restaurants.map((restaurant, i) => (
                  <Restaurant restaurant={restaurant} saved={true}
                    key={i} />
                ))
              ) : <p>No saved restaurants</p>
            }
          </div>
        </Tab>
      </Tabs>
    </>
  )
}
}
```

When all is complete, we'll keep our `api.js` file running and start our React app with `npm start` and our app is complete!

It's time to wrap up this chapter.

Summary

We've covered a *lot* of ground in this chapter. The power of JavaScript on both the frontend and the backend shows us that we can truly replace Python for many of our application needs. We've used a lot of React, but keep in mind that any frontend can be substituted here: Vue, Angular, and even frameworkless HTML, CSS, and JavaScript are at our disposal to create powerful web applications.

One of the things to note when using JavaScript and APIs is that there are cases when we need a middleware layer, for example, when saving files or accessing REST APIs with keys. Combining Express for powerful routing with a basic Node.js script to interact with an API is just the beginning of what we can accomplish with JavaScript and Node.js tied together.

In the next chapter, we'll explore webpack, a utility that allows us to logically combine and package our JavaScript application for deployment.

16
Enter Webpack

So, you now have beautiful frontend and backend code. Great! It looks so pretty sitting there on your laptop... so what's the next step? Publishing it to the world! It sounds easy, but when we have advanced JavaScript usage, such as with React, there are a couple more steps we might want to take to ensure our code is running at peak efficiency, all dependencies are resolved, and everything is compatible with modern technologies. Additionally, download size is a major consideration, so let's explore webpack, a tool to help mitigate these concerns.

We will cover the following points in this chapter:

- The need for bundling and modules
- Using webpack
- Deployment

Technical requirements

Be prepared to work with the code provided in the `Chapter-16` directory of the repository: `https://github.com/PacktPublishing/Hands-on-JavaScript-for-Python-Developers/tree/master/chapter-16`. As we'll be working with command-line tools, also have your terminal or command-line shell available. We'll need a modern browser and a local code editor.

The need for bundling and modules

Ideally, everything will work seamlessly on a website, without the need for any additional steps to be taken. You take your source files, drop them on a web server, and voilà: a site. However, this isn't always the case. For example, with React, we need to run `npm run build` to generate an output distribution directory for our project. We might also have other types of non-source files, such as SASS or TypeScript, which need to be converted into native file formats that the browser can understand.

So, what is a *module*? There's the concept of **modular programming**, which takes large programs and separates them by concern and encapsulation (scope) into smaller, contained chunks called modules. The ideas behind modular programming are many: scope, abstraction, logical design, testing, and debugging. Similarly, a bundle is a chunk of code that a browser can easily use, usually constructed from one or more modules.

Now here's the fun part: *we've already worked with modules!* Let's take a look at some Node.js code we wrote in Chapter 11, *What is Node.js?*:

```
const readline = require('readline')
const randomNumber = Math.ceil(Math.random() * 10)

const rl = readline.createInterface({
 input: process.stdin,
 output: process.stdout
});

askQuestion()

function askQuestion() {
 rl.question('Enter a number from 1 to 10:\n', (answer) => {
   evaluateAnswer(answer)
 })
}

function evaluateAnswer(guess) {
 if (parseInt(guess) === randomNumber) {
   console.log("Correct!\n")
   rl.close()
   process.exit(1)
 } else {
   console.log("Incorrect!")
   askQuestion()
 }
}
```

On that very first line, we're using a module called `readline`, which, if you recall from our program, will be used to take user input from the command line. We've also used them in React—any time we've needed to use `npm install`, we're using the concept of modules. So why is this important? Let's consider a standard `create-react-app` installation from scratch:

1. Use `npx` to create a new React project: `npx create-react-app sample-project`.
2. Navigate into the directory and install the dependencies: `cd sample-project ; npm install`.
3. Start the project with `npm start`.

If you remember, this gives us a very interesting start page:

Figure 16.1 – The React start page

What are we really getting when we run `npm install`? Let's take a look at our file structure:

```
.
├── README.md
├── package-lock.json
├── package.json
├── public
│   ├── favicon.ico
│   ├── index.html
│   ├── logo192.png
│   ├── logo512.png
```

```
|       ├──── manifest.json
|       └──── robots.txt
├──── src
|       ├──── App.css
|       ├──── App.js
|       ├──── App.test.js
|       ├──── index.css
|       ├──── index.js
|       ├──── logo.svg
|       ├──── serviceWorker.js
|       └──── setupTests.js
└──── yarn.lock
```

Simple enough so far. However, in this listing, I've purposely excluded the `node_modules` directory. This has 18 files. Try running this command at the root directory of our project, which does not exclude that directory: `tree`. Enjoy watching the flurry of lines—32,418 files! Where did all of those come from? Our friend, `npm install`!

package.json

Our project's structure is controlled, in part, by our `package.json` file to manage dependencies. Most bundlers, such as webpack, will leverage the information in this file to create our dependency graphs and bite-sized modules. Let's take a look at it:

package.json

```
{
  "name": "sample-project",
  "version": "0.1.0",
  "private": true,
  "dependencies": {
    "@testing-library/jest-dom": "^4.2.4",
    "@testing-library/react": "^9.3.2",
    "@testing-library/user-event": "^7.1.2",
    "react": "^16.13.1",
    "react-dom": "^16.13.1",
    "react-scripts": "3.4.1"
  },
  "scripts": {
    "start": "react-scripts start",
    "build": "react-scripts build",
    "test": "react-scripts test",
    "eject": "react-scripts eject"
  },
  "eslintConfig": {
```

```
    "extends": "react-app"
  },
  "browserslist": {
    "production": [
      ">0.2%",
      "not dead",
      "not op_mini all"
    ],
    "development": [
      "last 1 chrome version",
      "last 1 firefox version",
      "last 1 safari version"
    ]
  }
}
```

This is a standard, basic package file; it only contains six dependencies: half for testing and half for React. Now, here's the fun part: each of those dependencies in turn has its own dependencies, which is how we end up with 32,400 files in our `node_modules` directory alone. By using modules, we don't have to hand-construct or manage dependencies; we can follow the DRY principles and leverage existing code that others (or we) have written in the form of modules. As we discussed when comparing Python and Node.js, `npm install` is similar to `pip install` in that we use packages in our Python program with the `import` keyword, while we use `require` in Node.js.

When we use `npm install` to install a new package into our project, it will add an entry into `package.json`. This is a file where you'll want to be very careful if you make any edits. In general, you shouldn't need to make too many changes, and you especially should avoid making substantial changes to the dependencies. Leverage the `install` command for that.

Build pipelines

Let's take a look at what happens when we prepare our React project for deployment. Run `npm run build` and observe the output. You should see an output similar to the following:

```
Creating an optimized production build...
Compiled successfully.

File sizes after gzip:

  39.39 KB  build/static/js/2.deae54a5.chunk.js
  776 B     build/static/js/runtime-main.70500df8.js
```

```
   650 B      build/static/js/main.0fefaef6.chunk.js
   547 B      build/static/css/main.5f361e03.chunk.css

The project was built assuming it is hosted at /.
You can control this with the homepage field in your package.json.

The build folder is ready to be deployed.
You may serve it with a static server:

  yarn global add serve
  serve -s build

Find out more about deployment here:

  bit.ly/CRA-deploy
```

If you take a look inside your build directory, you'll see nicely minified JavaScript files, packaged for efficient deployment. Here's the fun part: `npm run build` *from* `create-react-app` *uses webpack under the hood!* The `create-react-app` setup handles these pieces for us. It's a bit tricky to modify the innards of the `create-react-app` webpack setup, so let's now look at using webpack directly beyond the use case of React.

Using webpack

Now, webpack is one of many modular tools that can be used in your program. Additionally, unlike React scripts, it has use outside of React: it can be used as a bundler for many different types of applications. To get our hands dirty, let's create a small, useless sample project:

1. Create a new directory and navigate into it: `mkdir webpack-example ; cd webpack-example`.
2. We'll be using NPM, so we need to initialize it. We'll also accept the defaults: `npm init -y`.
3. We have to then install webpack: `npm install webpack webpack-cli --save-dev`.

Note that we're using `--save-dev` here because we don't need webpack to be built into our production-level files. By using dev dependencies, we can help reduce our bundle size, a factor that can slow down applications if it bloats.

If you look in the `node_modules` directory here, you'll see that we've already installed over 3.5 thousand files from our dependencies. Our project is fairly boring as it stands: we don't have any content! Let's fix that and make some files, as follows:

src/index.html

```
<!DOCTYPE html>
<html lang="en">
<head>
 <meta charset="UTF-8">
 <meta name="viewport" content="width=device-width, initial-scale=1.0">
 <title>Webpack Example</title>
</head>
<body>
 <h1>Welcome to Webpack!</h1>
 <script src="index.js"></script>
</body>
</html>
```

src/index.js

```
console.log('hello')
```

So far, very exciting and useful, right? If you open our index page in a browser, you'll see what you would expect in the console. We'll now introduce webpack into the mix:

1. Change the `package.json scripts` node to the following:

```
"scripts": {
    "test": "echo \"Error: no test specified\" && exit 1",
    "dev": "webpack --mode development",
    "build": "webpack --mode production"
},
```

2. Run `npm run dev`. You should see an output like this:

```
> webpack --mode development

Hash: 21e0ae2cc4ae17d2754f
Version: webpack 4.43.0
Time: 53ms
Built at: 06/14/2020 1:37:27 PM
  Asset       Size  Chunks               Chunk Names
main.js  3.79 KiB    main  [emitted]  main
Entrypoint main = main.js
[./src/index.js] 20 bytes {main} [built]
```

Now go ahead and look in your newly created `dist` directory:

```
dist
└── main.js
```

If you open `main.js`, you'll find it looks *quite* different than our `index.js`! That's webpack doing some under-the-hood work to make our first steps into modularization.

Wait. We went from one line of code to 100. Why on earth is this better? Well, for this simple of an example it may not be, but humor me for a while longer. Let's try `npm run build` and compare the output: `main.js` is now one line, minified.

Taking a look at our `package.json` file, we'll see a few noteworthy pieces apart from the script node we manipulated:

```json
{
  "name": "webpack-example",
  "version": "1.0.0",
  "description": "",
  "main": "index.js",
  "scripts": {
    "test": "echo \"Error: no test specified\" && exit 1",
    "dev": "webpack --mode development",
    "build": "webpack --mode production"
  },
  "keywords": [],
  "author": "",
  "license": "ISC",
  "devDependencies": {
    "webpack": "^4.43.0",
    "webpack-cli": "^3.3.11"
  }
}
```

We see a `"main"` node that specifies an `index.js` to be used for our main entry point, or where webpack begins looking to catalog its dependencies.

There are three main concepts that are important to understand when using webpack:

- **Entry**: The place where webpack starts its work.
- **Output**: The place where webpack will spit out its finished product. If we look at the output of the preceding test, we'll see `main.js   3.79 KiB      main [emitted]   main`. Instead of the phrase "spits out," webpack much more elegantly defines it as "emitting" its bundles.

- **Loaders**: As mentioned, webpack can be used for a variety of different purposes; however, by default, webpack only really processes JavaScript and JSON files. So, we use *loaders* to do more work. We'll use one in a minute to manipulate the `index.html` file.

The concepts of mode and plugins are also important, though a bit more self-explanatory: mode, as we saw when we added our scripts to `package.json`, defines whether we want development, production, or "none" for our environment optimizations. There's more to mode than that, but for now, we won't go crazy—webpack is quite complex, so a surface-level understanding is a good place to begin. Plugins basically do what loaders can't. We'll keep it simple, though, and now we'll add a loader that understands HTML files. Brace yourself... the output isn't *quite* what you're thinking it will be:

1. Run `npm install html-loader --save-dev`.
2. We've now arrived at the point where we need a configuration file, so create `webpack.config.js`.
3. Enter this inside `webpack.config.js`:

```
module.exports = {
  module: {
    rules: [
      {
        test: /\.html$/i,
        loader: 'html-loader',
      },
    ],
  },
};
```

4. Modify `index.js`, as follows:

```
import html from './index.html'

console.log(html)
```

5. Modify `index.html` by changing the script tag as follows: `<script src="../dist/main.js"></script>`.
6. Re-run `npm run dev` and then open that page in a browser.

If we look at our console, we'll see our HTML! Woo! Everything's pretty much there, except our `<script>` tag says "`[Object object]`" in the `src`. Now you should be asking yourself: "what on earth did we just accomplish?".

As it turns out, a loader *isn't* what we want! It's a common mistake to work with loaders when you want plugins, and vice versa. Let's now unwind what we did and install an HTML plugin that *will* do what we expect: insert `index.html` into the `dist` directory with an optimized `main.js` file:

1. We don't actually want or need the HTML loader for this task: `npm uninstall html-loader`.

2. Install the right plugin: `npm install html-webpack-plugin --save-dev`.

3. Completely replace the contents of `webpack.config.js` with this configuration:

```
var HtmlWebpackPlugin = require('html-webpack-plugin');
var path = require('path');

module.exports = {
  entry: './src/index.js',
  output: {
    path: path.resolve(__dirname, './dist'),
    filename: 'index_bundle.js'
  },
  plugins: [new HtmlWebpackPlugin({
    template: './src/index.html'
  })]
};
```

4. Modify `index.js` back to its original one line: `console.log('hello')`.

5. Remove the `<script>` tag from `src/index.html`. It will be built for us.

6. Execute `npm run dev`.

7. Finally, open `dist/index.html` in a browser.

That should be more to your liking and what you would expect by using webpack. Now, however, this is a very basic example, so let's see whether we can do something fancier. Edit the files as follows:

src/index.html

```
<!DOCTYPE html>
<html lang="en">
<head>
 <meta charset="UTF-8">
 <meta name="viewport" content="width=device-width, initial-scale=1.0">
 <title>Webpack Example</title>
</head>
<body>
 <h1>Welcome to Webpack!</h1>
 <div id="container"></div>
```

```
</body>
</html>
```

src/index.js

```
import Highcharts from 'highcharts'

// Create the chart
Highcharts.chart('container', {
  chart: {
    type: 'bar'
  },
  title: {
    text: 'Fruit Consumption'
  },
  xAxis: {
    categories: ['Apples', 'Bananas', 'Oranges']
  },
  yAxis: {
    title: {
      text: 'Fruit eaten'
    }
  },
  series: [{
    name: 'Jane',
    data: [1, 0, 4]
  }, {
    name: 'John',
    data: [5, 7, 3]
  }]
});
```

For this example, we're using Highcharts, a charting library. This is their boilerplate example, taken directly from their site; I haven't done anything fancy with it except modified line 1 to `import Highcharts from 'highcharts'`. That implies we're going to use a module, so let's install it—`npm install highcharts`:

1. Add this script to your `package.json scripts` node: `"watch": "webpack -- watch -- mode development"`.
2. Run `npm run watch`.

3. Load `dist/index.html` in the browser:

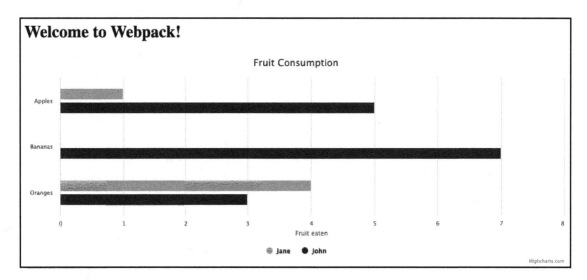

Figure 16.2 – Webpack with Highcharts

Much more interesting, don't you think? Also, take a moment to look at the `index_bundle.js` file and note the much larger file and minified code. If you edit a file in `src`, with `watch`, webpack will repackage the files on the fly for you. If you're using a live server that does hot reloads, such as with Visual Studio Code, it'll also refresh the page for you—handy for rapid development!

It's time to finally try what we've been building. Let's try building our project for deployment next.

Deploying our project

We've done a lot of development work so far, and now it's time to try a production build of our project. Run `npm run build` and, well, it's not quite so happy, is it? You should get some warnings like this:

```
WARNING in asset size limit: The following asset(s) exceed the recommended
size limit (244 KiB).
This can impact web performance.
Assets:
  index_bundle.js (263 KiB)
```

```
WARNING in entrypoint size limit: The following entrypoint(s) combined
asset size exceeds the recommended limit (244 KiB). This can impact web
performance.
Entrypoints:
  main (263 KiB)
      index_bundle.js


WARNING in webpack performance recommendations:
You can limit the size of your bundles by using import() or require.ensure
to lazy load some parts of your application.
For more info visit https://webpack.js.org/guides/code-splitting/
Child HtmlWebpackCompiler:
    1 asset
    Entrypoint HtmlWebpackPlugin_0 = __child-HtmlWebpackPlugin_0
    [0] ./node_modules/html-webpack-plugin/lib/loader.js!./src/index.html
522 bytes {0} [built]
```

So, what is this trying to tell us? Remember when I was talking about bundle size impacting performance? Let's try to optimize this so that we no longer get these messages. We'll investigate a few development techniques to do this.

Chunking

Simply put, chunking is the method of taking large files and splitting them apart into smaller chunks. We can do this part easily by adding this to our webpack.config.js file after our plugins node:

```
optimization: {
  splitChunks: {
    chunks: 'all',
  }
}
```

Now, go ahead and build; it'll be a *little* happier:

```
Built at: 06/14/2020 3:46:38 PM
                Asset       Size   Chunks                    Chunk Names
           index.html   321 bytes          [emitted]
       main.bundle.js   1.74 KiB        0  [emitted]         main
vendors~main.bundle.js    262 KiB        1  [emitted] [big]  vendors~main
Entrypoint main [big] = vendors~main.bundle.js main.bundle.js
```

It's still going to complain, unfortunately. We shaved 1.74 KB into a separate file, but we still have our `vendors` bundle sitting at 262 KB. If you look in `dist`, you'll now see the two `js` files as well as two `<script>` tags in the HTML.

The reason it's not splitting it further is that the vendor (Highcharts) bundle is pretty self-contained already, so we need to explore other ways to accomplish what we need. If we had a lot of our own code, however, it would probably be splitting it further into multiple chunks.

So, what's our next option? We tweak the optimization!

Try this one on for size:

```
optimization: {
    splitChunks: {
        chunks: 'async',
        minSize: 30000,
        maxSize: 244000,
        minChunks: 2,
        maxAsyncRequests: 6,
        maxInitialRequests: 4,
        automaticNameDelimiter: '~',
        cacheGroups: {
            defaultVendors: {
                test: /[\\/]node_modules[\\/]/,
                priority: -10
            },
            default: {
                minChunks: 2,
                priority: -20,
                reuseExistingChunk: true
            }
        }
    }
}
```

If you note, the options in this are much more explicit, including a maximum size for the chunks, reusing existing vendor chunks, and a minimum number of chunks. Let's try it.

No change, right?

Let's try something different: modify `index.js` to use promises and **webpack hints** to break the Highcharts dependency into its own bundle:

```
import( /* webpackChunkName: "highcharts" */ 'highcharts').then(({ default: Highcharts }) => {
    // Create the chart
```

```
Highcharts.chart('container', {
  chart: {
    type: 'bar'
  },
  title: {
    text: 'Fruit Consumption'
  },
  xAxis: {
    categories: ['Apples', 'Bananas', 'Oranges']
  },
  yAxis: {
    title: {
      text: 'Fruit eaten'
    }
  },
  series: [{
    name: 'Jane',
    data: [1, 0, 4]
  }, {
    name: 'John',
    data: [5, 7, 3]
  }
  ]
});
})
```

Our output from `npm run build` should now look more like this:

```
Version: webpack 4.43.0
Time: 610ms
Built at: 06/14/2020 4:38:41 PM
                      Asset       Size  Chunks                      Chunk
Names
highcharts~c19dcf7a.bundle.js   262 KiB      0  [emitted]  [big]
highcharts~c19dcf7a
                 index.html  284 bytes         [emitted]
     main~d1c01171.bundle.js   2.33 KiB      1  [emitted]
main~d1c01171
Entrypoint main = main~d1c01171.bundle.js
```

Well… that *still* didn't do what we wanted! While we have a separate chunk for Highcharts, it's still a large, monolithic file. So, what do we do?

Surrender

Raise the white flag. Admit defeat.

Almost.

Here's where each vendor package may differ and each import will be unique; what we want to do is try to find the *smallest chunk* of our vendor library that suits our needs. In this case, importing all of Highcharts is creating a massive file. However, let's take a look at `node_modules/highcharts`. Inside the `es-modules` directory, there's an interesting file: `highcharts.src.js`. This is a more modular file of what we want, so let's try importing it instead of the whole library at once:

```
import( /* webpackChunkName: "highcharts" */ 'highcharts/es-
modules/highcharts.src.js').then(({ default: Highcharts }) => {

  . . .
```

Now see what happens if we use `npm run build`:

```
Version: webpack 4.43.0
Time: 411ms
Built at: 06/14/2020 4:48:43 PM
                      Asset       Size  Chunks          Chunk Names
highcharts~47c7b5d6.bundle.js    170 KiB      0  [emitted]
highcharts~47c7b5d6
                 index.html  284 bytes         [emitted]
     main~d1c01171.bundle.js    2.33 KiB       1  [emitted]  main~d1c01171
Entrypoint main = main~d1c01171.bundle.js
```

Aha! Much nicer! So, in this case, the answer was obtuse. Highcharts bundling can be unwound in order to add only specific pieces of the code. This *will not* work in all cases, especially where the source is not included; however, this is a course of approach for us at this time: pare down the included packages to the smallest needed set. Remember when we selectively included parts of libraries in React? The same idea holds true here.

Deployment, finished

Now what do we do? All you really have to do is take the contents of your `dist` directory and put it on a web server for the world to see! Let your hard work be shown.

Summary

Webpack is our friend. It modularizes, minifies, chunks, and makes our code more efficient, as well as warning us when certain pieces aren't properly optimized. There are ways to silence these alerts, but in general, it's a good idea to listen to them and at least *try* to resolve them.

One burning question, though, that remains unanswered: doesn't increasing the number of files downloaded increase the load time? This is a common misconception that's hung around from the early days of the web: more files == more load time. The fact is, however, that multiple browsers can open many non-blocking streams simultaneously, allowing for a *more* efficient download than one huge file. Is this a solution for all multiple files? No: a CSS image sprite, for example, is still a more efficient use of image resources. For performance, we must toe a fine line on how to provide the best user experience, while combining that with the best developer experience. Entire books are written on this topic alone, so I won't try to give you all the answers. I'll just leave you with this:

Optimize, optimize, optimize.

In the next chapter, we'll deal with a very important topic for all parts of programming: security.

Section 4 - Communicating with Databases

4

The last part of our full-stack experience with JavaScript is the database layer. We'll be using NoSQL datastores, as they use JSON-like documents.

In this section, we will cover the following chapters:

17
Security and Keys

Security is no simple matter. It's important to keep security in mind when designing your applications from the beginning. For example, if you accidentally committed your keys to your repository, you'd have to do some trickery to either remove that from the repository's history or, more likely, you'd have to revoke those credentials and generate new ones.

We simply can't have our database credentials visible to the world in our frontend JavaScript, but there are ways for the frontend to work with databases. The first step is to implement the proper security and understand where we can put our credentials, both for the frontend and the backend.

The following topics will be covered in this chapter:

- Authentication versus authorization
- Using Firebase
- `.gitignore` and environment variables for credentials

Technical requirements

Be prepared to work with the code provided in the `Chapter-17` directory of the repository: `https://github.com/PacktPublishing/Hands-on-JavaScript-for-Python-Developers/tree/master/chapter-17`. As we'll be working with command-line tools, also have your terminal or command-line shell available. We'll need a modern browser and a local code editor.

Authentication versus authorization

As we begin our exploration of security with JavaScript, it's important to understand the vital difference between **authentication** and **authorization**. In a nutshell, *authentication* is a process whereby a system affirms and acknowledges that you are who you say you are. Think of going to the store to buy a bottle of wine. You may be asked to provide identification that proves you are of or above the legal consumption age of your locale. The clerk has *authenticated* you with your photo ID to say that *yes, you are **you** because I, the clerk, have matched your face to the photo in the I.D.* A second case is when you fly on an airline. When you pass through security, they're also going to check your ID for the same reason: are you who you say you are?

These two use cases end, however, with *authorization*. Authorization says: *I know you are who you say you are.* Now, are you allowed to do what you want? In our wine example, if you are above the age of 21 in the United States, or 18 in most other parts of the world, you are *authorized* to consume alcoholic beverages. Now, the security agent at the airport does not really care about your age for any real reason; they are only concerned about whether you are who you say you are and whether you have a valid ticket for the flight you're about to board. You are then *authorized* to enter the secure area of the airport and board your flight.

Let's continue our airline example a step further. In today's age of enhanced security for travel, the authentication and authorization process neither begins nor ends with the security agent. The process looks more like this if you book a commercial airline ticket online:

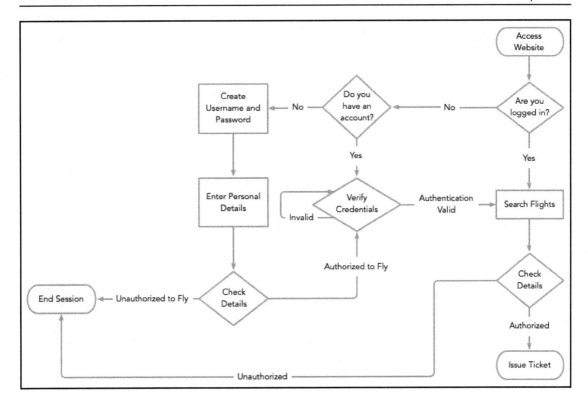

Figure 17.1 – Airline website authentication and authorization

When using an airline's website, you may have an account and be *authorized* to proceed to log in, or you may already be logged in and *authorized* to search flights. If you are logged out, you must *authenticate* to search for flights. To book a flight, you may have to have certain details, such as a visa, in order to be *authorized* to book that flight. You may also be on a watchlist or blacklist for traveling to a country, so your journey would end before it even began. There are so many steps, but many of these happen behind the scenes; for example, you may not know that when you entered your name to book a ticket, your name was searched against global records to see whether you are authorized to fly. Your visa number may have been cross-referenced to see whether you are authorized to fly to that country.

Just as you need to be authenticated and authorized to fly, your web application should also be designed to allow authentication and authorization. Consider our restaurant finder application from `Chapter 15`, *Combining Node.js with the Frontend*, that allowed us to search for and save different restaurants in Firebase:

Figure 17.2 – Our restaurant app

If you remember, we started our Firebase app with *open permissions* in the **Real-Time Database** section:

```
Simulator

⚠ Your security rules are defined as public, so anyone can steal, modify, or delete data in your database    Learn more  Dismiss

1 ▾  {
2 ▾     "rules": {
3          ".read": true,
4          ".write": true
5       }
6  }
```

Figure 17.3 – Our Firebase security rules

This is obviously *not a good idea* for a production website. So, to mitigate this, let's return to Firebase and set up some authentication and authorization!

Using Firebase

For ease of use, I've replicated our restaurant finder app in the `Chapter-17` directory in the GitHub repository. Don't forget to include your own environment variables in the `.env` file from `Chapter 15`, *Combining Node.js with the Frontend*, restaurant finder. Take a moment to get this set up and working before we move on.

The next thing we'll need to do is go to Firebase and configure it to use authentication. In the Firebase console, access the **Authentication** section and set up a sign-in method; for example, you can set up Google authentication. There's a list of methods you can use here, so go ahead and add one or more.

Next, we're going to set our rules in the **Real-Time Database** section, as follows:

```
{
  "rules": {
    "restaurants": {
      "$uid": {
        ".write": "auth != null && auth.uid == $uid",
        ".read": "auth != null && auth.uid == $uid"
      }
    }
  }
}
```

What we're saying here is that the user is allowed to read and write from the `restaurants/<user id>` section of your database if the authenticated data is not `null` and if the user ID matches the user ID of the location in the database to which you are attempting to write and read.

Now that our rules are set up, let's try to save a restaurant:

1. Start the app by executing `npm start` in the root directory and access `http://localhost:3000`.
2. Search for a restaurant.
3. Attempt to save the restaurant.
4. Witness an epic fail.

What you should see is an error screen that looks something like this:

```
PERMISSION_DENIED: Permission denied                              ✕

(anonymous function)
src/core/Repo.ts:624

  621 |    message += ': ' + errorReason;
  622 |  }
  623 |
> 624 |  const error = new Error(message);
  625 |  // eslint-disable-next-line @typescript-eslint/no-explicit-any
  626 |  (error as any).code = code;
  627 |  callback(error);
```

Figure 17.4 – Error, error!

Additionally, if we go to our developer tools and inspect the **Network** tab's **WS** tab (**WS** for **WebSockets**, which is how Firebase communicates), we might see something like this:

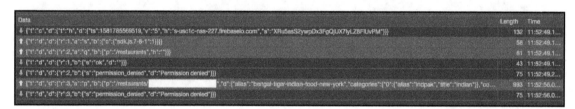

Figure 17.5 – WebSockets communication inspector

Great! We've now proven that our Firebase rules work and will not allow saving to `/restaurants/<user_id>` because we are not authenticated. It's time to set that up.

The first thing we're going to do is change our `App.js` script a bit. There are a few different conventions when writing React, and we're going to continue moving forward with the class-based method. Here's what our `App.js` script will look like:

```
import React from 'react'
import cookie from "react-cookies"

import Finder from './components/finder/Finder'
import SignIn from './components/signIn/SignIn'

import './App.css'
```

```
export default class App extends React.Component {
  constructor() {
    super()

    this.state = {
      user: cookie.load("username")
    }
    this.setUser = this.setUser.bind(this)
  }

  setUser(user) {
    this.setState({
      user: user
    })

    cookie.save("username", user)
  }

  render() {
    const { user } = this.state
    return (
      <div className="App">
        { (user) ? <Finder user={user} /> : <SignIn setUser={this.setUser}
        /> }
      </div>
    )
  }
}
```

The first piece to notice is that we've included a new npm module: react-cookies. While cookies are easy to read from a browser, there are modules that make it just a little bit easier. When we retrieve the user's ID, we're going to store it in a cookie so that the browser remembers that the user is authenticated.

Why do we need to use a cookie? If you remember, the web is inherently *stateless*, so cookies are one means of transferring information from one part of an application to another and from session to session. It's a basic example, but it's important to remember not to store any sensitive information in a cookie; a token or username is probably the most that you'd want to put in one for an auth workflow.

We've also introduced a new component, SignIn, which is conditionally rendered if the user variable doesn't exist—that is, if the user isn't logged in. Let's take a look at that component:

```
import React from 'react'
import { Button } from 'react-bootstrap'
import * as firebase from 'firebase'
```

```
const provider = new firebase.auth.GoogleAuthProvider()

export default class SignIn extends React.Component {
  constructor() {
    super()

    this.login = this.login.bind(this)
  }

  login() {
    const self = this

    firebase.auth().signInWithPopup(provider).then(function (result) {
      // This gives you a Google Access Token. You can use it to access the
      // Google API.
      var token = result.credential.accessToken;
      // The signed-in user info.
      self.props.setUser(result.user);
      // ...
    }).catch(function (error) {
      // Handle Errors here.
      var errorCode = error.code;
      var errorMessage = error.message;
      // The email of the user's account used.
      var email = error.email;
      // The firebase.auth.AuthCredential type that was used.
      var credential = error.credential;
      // ...
    });
  }
  render() {
    return <Button onClick={this.login}>Sign In</Button>
  }
}
```

There are two things to note here:

- We're using `GoogleAuthProvider` for our `SignIn` mechanism. If you chose a different authentication method when setting up Firebase, this provider may be different, but the rest of the code should be the same or similar.
- The `signInWithPopup` method is copied almost directly from the Firebase documentation. The only change made here is to create the `self` variable so that we can maintain the scope to `this` inside another method.

When this is rendered, it'll be a simple button stating **Sign In** if the user isn't already logged in. It will activate a popup to log in with your Google account, and then proceed as before. Not so scary, right?

Next, we need to deal with our user. Did you notice in App.js that we're passing the user prop to Finder? That'll make it easy to pass a reference to our user in our basic application, as follows in Finder.jsx:

```
getRestaurants() {
    const { user } = this.props

    Database.ref(`/restaurants/${user.uid}`).on('value', (snapshot) => {
      const restaurants = []

      const data = snapshot.val()

      for(let restaurant in data) {
        restaurants.push(data[restaurant])
      }
      this.setState({
        restaurants: restaurants
      })
    })
  }
```

This is the only method that's changed in this instance, and if you look closely, the change is to destructure user from this.props and use it in our database reference. If you remember our security rules, we've had to change our database structure a bit to accommodate easy *authorization* of our authenticated user:

```
{
  "rules": {
    "restaurants": {
      "$uid": {
        ".write": "auth != null && auth.uid == $uid",
        ".read": "auth != null && auth.uid == $uid"
      }
    }
  }
}
```

What we stated in our security rules is that the node of the format `restaurants.$uid` is where we'll store each individual user's restaurants. Our Firebase structure now looks something like this:

Figure 17.6 – An example of how our Firebase structure could look

In this construction, we see the `TT8PYnjX6FP1YikssoHnINIpukZ2` node inside `restaurants`. That's the **uid (user ID)** of the authenticated user, and within that node, we find the user's saved restaurants.

This database structure is simple but provides easy authorization. Our rules state "give the user TT8 permission to see and alter the data within their own node and nothing more."

We've discussed our .env variables a bit previously, so let's take a bit of a deeper look into them. We'll be deploying our app to Heroku to create a publicly visible website.

.gitignore and environment variables for credentials

As we've been working with .env files, I've made it a point to note that these files should *never* be committed to the repository. In fact, a good practice is to add an entry to your .gitignore file before you create any sensitive files to ensure you never accidentally commit your credentials. Even if you later delete it from your repository, the file history is maintained and you'll have to invalidate (or *cycle*) those keys so that they are not exposed in history.

While a full section on Git is beyond the scope of our work here, let's take a look at an example of a .gitignore file:

```
# See https://help.github.com/articles/ignoring-files/ for more about
ignoring files.

# dependencies
/node_modules
/.pnp
.pnp.js

# testing
/coverage

# production
/build

# misc
.DS_Store
.env*

npm-debug.log*
yarn-debug.log*
yarn-error.log*
```

Several of these are entries created by the create-react-app scaffold. Note specifically .env*. The asterisk (or *star*, or *splat*) is a regular expression wildcard to specify that any file that starts with .env is ignored. You can have .env.prod and it will be ignored. **Be sure to ignore your credential files!**

I also like to change `/node_modules` to `*node_modules*` in case you have subdirectories with their own node modules.

Storing variables in `.env` files is convenient, but there are also in-memory environment variables that can be created. To demonstrate this functionality, we're going to deploy our project to Heroku, a cloud application platform. Let's get set up:

1. Create a new account at `https://heroku.com`.
2. Install the Heroku **Command-Line Interface (CLI)** as per the documentation provided. Be sure to follow the login instructions as well.
3. Initialize a new repository in the restaurant finder directory: `git init`.
4. Execute `heroku create --ssh-git`. It'll provide the Git URL of your Heroku endpoint, as well as the `https://` URL. Go ahead and access the HTTPS URL. You should see a welcome message:

Heroku | Welcome to your new app!

Refer to the <u>documentation</u> if you need help deploying.

Figure 17.7 – Hooray! We have a blank Heroku application!

We can now continue with organizing our application's logic.

Reorganizing our application

The next thing we're going to do that's different than in `Chapter 15`, *Combining Node.js with the Frontend*, is reorganize our files just a touch. This isn't completely necessary, but it provides a nice logical distinction between the frontend and backend, which is especially useful when deploying production-level code. There's an additional semantic difference between our previous application and what we're going to create here: we're not going to serve a running development React application, but rather a static production build.

If you recall, our previous restaurant had a structure that looked like this:

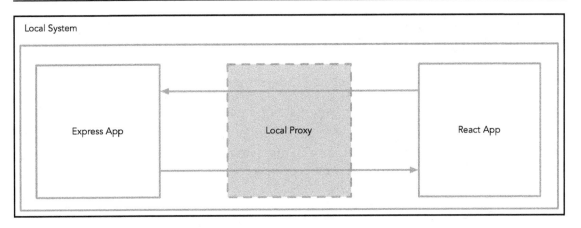

Figure 17.8 – Proxy versus apps, explained.

We were actually using the React app as the web server and proxying through it to the Express backend in order to use the Yelp API. Now, however, we're going to use Express as the main web server and serve a production-level build of our React application.

Our application logic previously looked as follows:

```
IF NOT a React page,
  Serve from proxy
ELSE
  Serve React
```

We're going to turn that logic on its head and state the following:

```
IF NOT an Express route,
  Serve from static build
ELSE
  Serve API
```

Here's what to do:

1. Create a new `client` directory.
2. Delete the `yarn.lock` file if you still have it. We're going to focus on using NPM instead of `yarn`.
3. Move all of the files into the client directory, except for the API directory.
4. Next, we're going to make a new `package.json` at our root level: `npm install dotenv express yelp-fusion`.

If you noticed, we also installed Express, which we didn't do before. We're going to use this to make routing our requests much easier.

In our `package.json`, at the *root* level, add these scripts:

```
"postinstall": "cd client && npm install && npm run build",
"start": "node api/api.js"
```

Since we're dealing with Heroku, we can also remove the `proxy` line from `package.json`, as everything will be running on the same server and will not need a proxy. Now, how about the `postinstall` line in our `package.json`? What we're going to do is make a *production-ready* build of our application. `create-react-app` gives us that functionality for free with the `npm run build` script. When we deploy to Heroku, it will run `npm install` and then, `postinstall`, to create a production build of our React application.

We're now ready to add a new piece of metadata to our project so that Heroku will serve up our application: the **Procfile**.

A Procfile will tell Heroku what to do with our code. Your Procfile will look like this:

```
web: npm start
```

In essence, all it's doing is telling Heroku where to begin the program: run `npm start`.

Our directory structure should now look like this:

```
.
├──── Procfile
├──── api
│     └──── api.js
├──── client
│     ├──── README.md
│     ├──── package-lock.json
│     ├──── package.json
│     ├──── public
│     └──── src
├──── package-lock.json
└──── package.json
```

Our next important step is to modify our `api.js` file, as follows:

```
const yelp = require('yelp-fusion');
const express = require('express');
const path = require('path');

const app = express();
```

```
require('dotenv').config();

const PORT = process.env.PORT || 3000;

const client = yelp.client(process.env.YELP_API_Key);
```

So far, this looks pretty similar to before, except for the addition of Express. But take a look at the next line:

```
app.use(express.static(path.join(__dirname, '../client/build')));
```

Aha! Here's our secret sauce: this line states to use the `client/build` directory as static assets as opposed to Node.js code.

Continuing on, we're defining our Express routing to handle requests of the format `/search`:

```
app.get('/search', (req, res) => {
 const { lat, lng, value } = req.query

 client.search({
   term: value,
   latitude: lat,
   longitude: lng,
   categories: 'Restaurants'
 }).then(response => {
   res.statusCode = 200;
   res.setHeader('Content-Type', 'application/json');
   res.setHeader('Access-Control-Allow-Origin', '*');

   res.write(response.body);
   res.end();
 })
   .catch(e => {
     console.error('error', e)
   })
});
```

For the next part of our secret sauce, if the route does *not* match `/search`, send it along to the static React build:

```
app.get('*', (req, res) => {
 res.sendFile(path.join(__dirname + '../client/build/index.html'));
});

app.listen(PORT, () => console.log(`Server listening on port ${PORT}`));
```

Add everything to your Git repository: `git add`. Now you can execute `git status` to be sure that your `.env` file is not *included*.

Next, commit your code: `git commit -m "Initial commit`. If you need some help with Git, the Heroku documentation provides references. Next, deploy to Heroku: `git push heroku master`. This will take a little while, as Heroku is going to not only deploy your code with Git but also create a production build of your code.

Access the URL provided by the build script and hopefully, you'll see a fantastic error message:

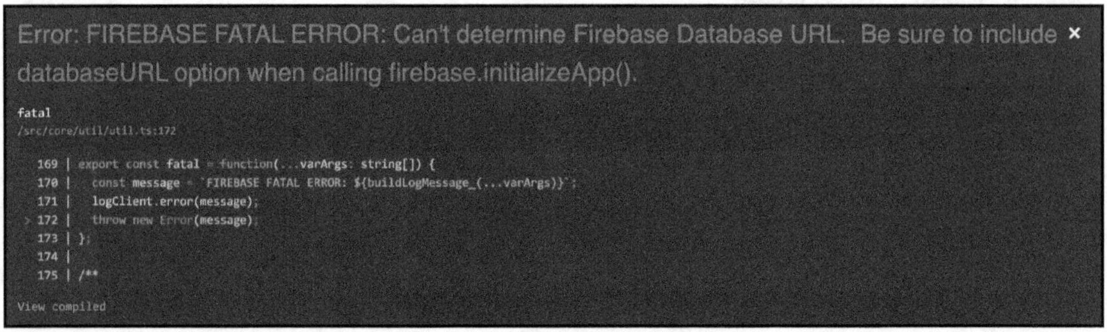

```
Error: FIREBASE FATAL ERROR: Can't determine Firebase Database URL. Be sure to include ✕
databaseURL option when calling firebase.initializeApp().

fatal
/src/core/util/util.ts:172

  169 |  export const fatal = function(...varArgs: string[]) {
  170 |    const message = `FIREBASE FATAL ERROR: ${buildLogMessage_(...varArgs)}`;
  171 |    logClient.error(message);
> 172 |    throw new Error(message);
  173 |  };
  174 |
  175 |  /**

View compiled
```

Figure 17.9 – Oh no! An error! Actually it's not a bad thing!

Great! What this is telling us is that the application is running, but we don't have some important pieces: our environment variables. Execute `heroku config:set <entry>` for each entry in your `.env` files (both the root and in `client`).

When you refresh the page, you'll see the **Sign In** button. However, if you click it, nothing will happen. It may spawn a popup for a second, but it won't bring up the authentication window. We need to go back to the Firebase console to add our Firebase URL as an *authorized* URL.

In the Firebase console, navigate to the authentication section and enter your Heroku URL to the **Authorized domains** section. Return to your Heroku app, refresh, and lo and behold! The authentication panel works. If you go to **Saved!**, you'll even see your saved restaurants.

That wasn't so bad! Heroku's method of storing environment variables doesn't vary too much from our `.env` file, but it handles it for us without needing to do much work. However, there's one last piece we'll need to configure: our search *does not work*. If you look at the console error messages, you should see a note indicating that a connection to `localhost:3000` is refused. We need to take one last step to abstract our code away from using `localhost`.

In `src/components/search/Search.jsx`, you may recognize this method:

```
search(event) {
    const { lng, lat, val } = this.state

fetch(`http://localhost:3000/businesses/search?value=${val}&lat=${lat}&lng=
${lng}`)
        .then(data => data.json())
        .then(data => this.handleSearchResults(data))
  }
```

OK! We've hardcoded our `fetch` call to `localhost` and our proxied path. Let's change it to the following:

```
fetch(`/search?value=${val}&lat=${lat}&lng=${lng}`)
```

Commit your changes and push to Heroku again. As you're developing, you can also use `heroku local web` to spawn a browser and test your changes without committing and deploying.

With any luck, you should have a fully functional front-to-back application with credentials secured in Heroku environment variables! Congratulations!

Summary

In this chapter, we learned about authentication, authorization, and the difference between the two. Remember that it's usually not enough to only do one or the other: most applications that need credentials need a combination of both.

Firebase is a useful cloud storage database that you can use with existing login systems and can not only be useful as a development resource but can also scale to production-level usage. Lastly, remember these points: because JavaScript is client-side, we have to protect sensitive information in different manners than a purely backend application:

1. Authenticate and authorize to determine who can use which resources.
2. Separate our sensitive data from our public data.
3. **Never commit keys and sensitive data to a repository!**

It's up to all of us to be good digital citizens, but there are bad actors out there. Protect yourself and your code!

In the next chapter, we'll be tying together Node.js and MongoDB to persist our data. We'll be revisiting our starship game, but this time with persistent storage.

18
Node.js and MongoDB

You may have heard of the **MEAN** stack: MongoDB, Express, Angular, and Node.js, or the **MERN** stack: MongoDB, Express, React, and Node.js. The missing piece that we have yet to discuss is MongoDB. Let's explore how this NoSQL database can be used directly from Express. We'll be constructing the next iteration of our starship game that we started in `Chapter 13`, *Using Express*, except this time using MongoDB and incorporating a bit of testing!

We will cover the following topics in this chapter:

- Using MongoDB
- Testing with Jest
- Storing and retrieving data
- Wiring your API together

Technical requirements

Be prepared to work with the code provided in the `chapter-18` directory of the repository: `https://github.com/PacktPublishing/Hands-on-JavaScript-for-Python-Developers/tree/master/chapter-18`. As we'll be working with command-line tools, also make sure to have your Terminal or command-line shell available. We'll need a modern browser and a local code editor.

Using MongoDB

The base premise behind MongoDB that makes it different from other types of structured key/value pair databases is that it's *schemaless*: you can insert arbitrary **documents** of unstructured data without concern for what another entry in the database looks like. A document in NoSQL parlance is something already familiar to us: a JavaScript object!

Here's a document:

```
{
 "first_name": "Sonyl",
 "last_name": "Nagale",
 "role": "author",
 "mood": "accomplished"
}
```

We can see that it's a basic JavaScript object; more specifically, it's JSON, which means it can also support nested data. Here's an example:

```
{
 "first_name": "Sonyl",
 "last_name": "Nagale",
 "role": "author",
 "mood": "accomplished",
 "tasks": {
  "write": {
   "status": "incomplete"
  },
  "cook": {
   "meal": "carne asada"
  },
  "read": {
   "book": "Le Petit Prince"
  },
  "sleep": {
   "time": "8"
  }
 },
 "favorite_foods": {
  "mexican": ["enchiladas", "burritos", "quesadillas"],
  "indian": ["saag paneer", "murgh makhani", "kulfi"]
 }
}
```

So how does this differ from MySQL? Consider this MySQL schema:

Field	Type		Length	Unsigned	Zerofill	Binary	Allow Null	Key	Default	Extra	Encoding	Collation	Comment
id	INT	⌄	11	✓	☐			PRI		auto_increment ⌄		⌄	⌄
admin_role_id	INT	⌄	11	☐	☐		✓		NULL	None ⌄			
first_name	VARCHAR	⌄	50			☐	✓		NULL	None ⌄	UTF-8 Unicode ⌄	utf8_unicode_ci ⌄	
last_name	VARCHAR	⌄	50			☐	✓		NULL	None ⌄	UTF-8 Unicode ⌄	utf8_unicode_ci ⌄	
username	VARCHAR	⌄	50			☐	✓		NULL	None ⌄	UTF-8 Unicode ⌄	utf8_unicode_ci ⌄	
email	VARCHAR	⌄	100			☐	✓	M...	NULL	None ⌄	UTF-8 Unicode ⌄	utf8_unicode_ci ⌄	
phone	VARCHAR	⌄	100			☐	✓		NULL	None ⌄	UTF-8 Unicode ⌄	utf8_unicode_ci ⌄	
password	VARCHAR	⌄	255			☐	✓	M...	NULL	None ⌄	UTF-8 Unicode ⌄	utf8_unicode_ci ⌄	
avatar	VARCHAR	⌄	100			☐	✓		NULL	None ⌄	UTF-8 Unicode ⌄	utf8_unicode_ci ⌄	
admin_role	ENUM	⌄	'admin','sub_admin'				✓	M...	NULL	None ⌄	UTF-8 Unicode ⌄	utf8_unicode_ci ⌄	
status	ENUM	⌄	'active','inactive',...				✓	M...	NULL	None ⌄	UTF-8 Unicode ⌄	utf8_unicode_ci ⌄	
last_login	DATETIME	⌄					✓		NULL	None ⌄		⌄	⌄
secret_key	VARCHAR	⌄	255			☐	✓		NULL	None ⌄	UTF-8 Unicode ⌄	utf8_unicode_ci ⌄	
last_login_ip	VARCHAR	⌄	50			☐	✓		NULL	None ⌄	UTF-8 Unicode ⌄	utf8_unicode_ci ⌄	
sidebar_status	ENUM	⌄	'open','close'				✓		open	None ⌄	UTF-8 Unicode ⌄	utf8_unicode_ci ⌄	
created	DATETIME	⌄					✓		NULL	None ⌄		⌄	
modified	DATETIME	⌄					✓		NULL	None ⌄		⌄	

Figure 18.1 – An example of a MySQL database table structure

If you're familiar with SQL databases, you know that each field type in a database table must be specifically typed. When retrieving from a SQL-type database, we use **Structured Query Language (SQL)**. Just as our tables are structured, so are our queries.

We need to create our database tables before using them, and in SQL it's advised not to change their structure once created without doing some additional cleanup work. Here's how we would create our preceding table:

```
CREATE TABLE `admins` (
 `id` int(11) unsigned NOT NULL AUTO_INCREMENT,
 `admin_role_id` int(11) DEFAULT NULL,
 `first_name` varchar(50) COLLATE utf8_unicode_ci DEFAULT NULL,
 `last_name` varchar(50) COLLATE utf8_unicode_ci DEFAULT NULL,
 `username` varchar(50) COLLATE utf8_unicode_ci DEFAULT NULL,
 `email` varchar(100) COLLATE utf8_unicode_ci DEFAULT NULL,
 `phone` varchar(100) COLLATE utf8_unicode_ci DEFAULT NULL,
 `password` varchar(255) COLLATE utf8_unicode_ci DEFAULT NULL,
 `avatar` varchar(100) COLLATE utf8_unicode_ci DEFAULT NULL,
 `admin_role` enum('admin','sub_admin') COLLATE utf8_unicode_ci DEFAULT
 NULL,
 `status` enum('active','inactive','deleted') COLLATE utf8_unicode_ci
 DEFAULT NULL,
 `last_login` datetime DEFAULT NULL,
 `secret_key` varchar(255) COLLATE utf8_unicode_ci DEFAULT NULL,
 `last_login_ip` varchar(50) COLLATE utf8_unicode_ci DEFAULT NULL,
 `sidebar_status` enum('open','close') COLLATE utf8_unicode_ci DEFAULT
 'open',
 `created` datetime DEFAULT NULL,
 `modified` datetime DEFAULT NULL,
 PRIMARY KEY (`id`),
```

```
    KEY `email` (`email`),
    KEY `password` (`password`),
    KEY `admin_role` (`admin_role`),
    KEY `status` (`status`)
) ENGINE=InnoDB DEFAULT CHARSET=utf8 COLLATE=utf8_unicode_ci;
```

Now, for MongoDB, we *won't* be constructing tables with predefined datatypes and lengths. Instead, we will insert JSON blobs into our database as **documents**. The idea behind MongoDB is very similar to when we used Firebase in Chapter 17, *Security and Keys*, inserting JSON and querying it, even with multiple nested JSON objects versus storing, cross-joining, and querying multiple tables.

Imagine that we have the following two documents:

```
{
  "first_name": "Sonyl",
  "last_name": "Nagale",
  "admin_role": "admin",
  "status": "active"
},
{
  "first_name": "Jean-Luc",
  "last_name": "Picard",
  "admin_role": "admin",
  "status": "inactive"
}
```

How do we insert them into our database? This would be with MySQL:

```
INSERT INTO
    admins(first_name, last_name, admin_role, status)
    VALUES
    ('Sonyl', 'Nagale', 'admin', 'active'),
    ('Jean-Luc', 'Picard', 'admin', 'inactive')
```

The answer with MongoDB is actually much easier than SQL, because we can place arrays easily and not have to worry about datatypes or ordering the data! We can just shove in the document without worrying about anything else, which is more likely to be how we receive it from the frontend:

```
db.admins.insertMany([
{
  "first_name": "Sonyl",
  "last_name": "Nagale",
  "admin_role": "admin",
  "status": "active"
},
```

```
{
  "first_name": "Jean-Luc",
  "last_name": "Picard",
  "admin_role": "admin",
  "status": "inactive"
}]
)
```

Now, for example, to get all active administrators from the preceding `admins` table, we would write something like this in MySQL:

```
SELECT
  first_name, last_name
FROM
  admins
WHERE
  admin_role = "admin"
AND
  status = "active"
```

The `first_name` and `last_name` fields are pre-defined as type VARCHAR (variable characters) with a maximum length of 50 characters. `admin_role` and `status` are ENUM (enumerated types) with predefined possible values (like a dropdown selection list on a site). However, here's how we would construct our query in MongoDB:

```
db.admins.find({ status: 'active', admin_role: 'admin'}, { first_name: 1,
last_name: 1})
```

We won't go *too* deep into MongoDB syntax here as it's a bit out of scope for this book and we'll only be using simple queries. With that being said, we should understand a bit more before we get started.

Here is the list of mongo commands we'll use while making our game:

- `find`
- `findOne`
- `insertOne`
- `updateOne`
- `updateMany`

Fairly manageable, right? We can break down many MongoDB commands into the following general syntactical structure:

```
<dbHandle>.<collectionName>.<method>(query, projection)
```

Here, `query` and `projection` are objects that dictate our usage of MongoDB. For example, in our preceding statement, `{ status: 'active', admin_role: 'admin' }` is our query to specify that we want those fields to equal those values. The projection in this example specifies what we want to return.

Let's dive into our project.

Getting started

The first thing we can do is download MongoDB Community Server from `https://MongoDBdb.com`. When you have it installed, navigate to the `chapter-18/starships` directory from our GitHub repository and let's try to get it started:

```
npm install
mkdir -p data/MongoDB
mongod --dbpath data/MongoDB
```

If you have everything installed correctly, you should see a flurry of notification messages, ending with one that says something similar to `[initandlisten] waiting for connections on port 27017`. If all does *not* go as planned, spend some time to ensure your installation is working properly. A useful tool is MongoDB Compass, a GUI tool for connecting to MongoDB. Be sure to check permissions and that the appropriate ports are open, as we'll use port `27017` (MongoDB's default port) for connections.

This chapter will be a lab exercise to take our starship game to the next level. Here's what we'll be building:

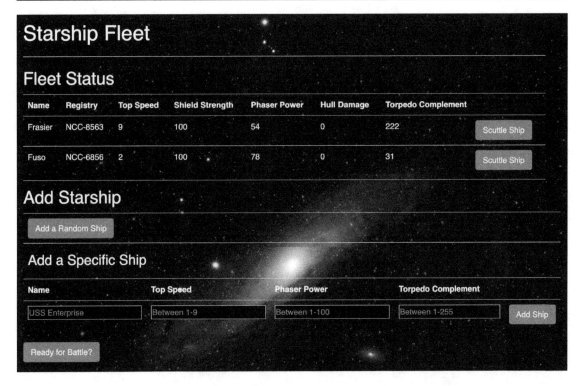

Figure 18.2 – Creating our fleet

Then, we're going to wire it to MongoDB and actually execute gameplay in this interface:

Name	Registry	Top Speed	Shield Strength	Phaser Power	Torpedo Complement	Hull Damage	Attack		
Frasier	NCC-8563	9	0	54	222	89	Fire at: Cube-7089 ⇕	Fire Phasers!	Fire Torpedo!
Fuso	NCC-6856	2	100	78	31	0	Fire at: Cube-7089 ⇕	Fire Phasers!	Fire Torpedo!

Name	Registry	Top Speed	Shield Strength	Phaser Power	Torpedo Complement	Hull Damage	Attack		
Borg Cube	Cube-7089	1	100	53	63	0	Fire at: NCC-8563 ⇕	Fire Phasers!	Fire Torpedo!
Borg Cube	Cube-3748	5	100	22	66	0	Fire at: NCC-8563 ⇕	Fire Phasers!	Fire Torpedo!

Figure 18.3 – Attack the enemy!

We'll be using a simplified version of MERN and using vanilla JavaScript instead of React, relying on Express to render our HTML in a less controlled manner than React. Maybe *JEMN stack* would be a good name?

Before we get to writing actual code, let's examine the setup of the project and get started with testing!

Testing with Jest

In the `starships` directory, you'll find the completed game. Let's dissect it.

Here's the directory listing:

```
.
├── README.md
├── app.js
├── bin
│   └── www
├── controllers
│   └── ships.js
├── jest-MongoDBdb-config.js
├── jest.config.js
├── models
│   ├── MongoDB.js
│   ├── setup.js
│   └── ships.js
├── package-lock.json
├── package.json
├── public
│   ├── images
│   │   └── bg.jpg
│   ├── javascripts
│   │   ├── index.js
│   │   └── play.js
│   └── stylesheets
│       ├── micromodal.css
│       └── style.css
├── routes
│   ├── enemy.js
│   ├── index.js
│   ├── play.js
│   ├── ships.js
│   └── users.js
├── tests
│   ├── setup.model.test.js
│   ├── ships.controller.test.js
```

```
|       └──── ships.model.test.js
└──── views
      ├──── enemy.hbs
      ├──── error.hbs
      ├──── index.hbs
      ├──── layout.hbs
      └──── play.hbs
```

We're going to take a little bit of a different approach here than our other projects and implement a very light cycle of **Test-Driven Development** (**TDD**). TDD is the practice of writing tests that *fail*, before writing your code that works. While we're not implementing true TDD, the idea of guiding our thought process using testing is what we'll be doing.

We'll be using Jest as our testing framework. Let's look at the steps:

1. Inside the `tests` directory, create a new file named `test.test.js`. The first `test` is the name for our test suite, and the convention of ending files in `.test.js` indicates to the testing framework that this is a test suite to execute. Inside the file, create this test script:

```
describe('test', () => {
  it('should return true', () => {
    expect(1).toEqual(1)
  });
});
```

2. Run the test with `node_modules/.bin/jest test.test.js` (be sure you've already run `npm install`!). You'll get output from the test suite similar to the following:

```
$ node_modules/.bin/jest test.test.js
 PASS  tests/test.test.js
   test
     ✓ should return true (2ms)

Test Suites: 1 passed, 1 total
Tests:       1 passed, 1 total
Snapshots:   0 total
Time:        0.711s, estimated 1s
Ran all test suites matching /test.test.js/i.
```

We've just written our first test suite! It simply says "I expect 1 to equal 1. If it does, pass the test. If not, fail the test." Pretty powerful for five lines of code, right? OK, maybe not, but this will provide us with the scaffolding for all of our other tests.

3. Let's look at the MongoDB model: `models/mongo.js`:

```
const MongoClient = require('mongodb').MongoClient;
const client = new MongoClient("mongodb://127.0.0.1:27017", {
useNewUrlParser: true, useUnifiedTopology: true });

let db;
```

4. So far, we're just setting up our MongoDB connection. Make sure you still have your MongoDB connection running now:

```
const connectDB = async (test = '') => {
  if (db) {
    return db;
  }

  try {
    await client.connect();
    db = client.db(`starships${test}`);
  } catch (err) {
    console.error(err);
  }

  return db;
}
```

5. As with all good database connection code, we're executing our code in a *try/catch* block to be sure that our connection is made correctly:

```
const getDB = () => db

const disconnectDB = () => client.close()

module.exports = { connectDB, getDB, disconnectDB }
```

Sneak peek: we're going to be using this `MongoDB.js` file from our tests and models. The `module.exports` line specifies which functions are exported from this file and exposed to other parts of our program. We'll be using this export directive consistently throughout our program: when we want to expose a method, we'll use an export.

6. Return to `test.test.js` and include our MongoDB model at the beginning of the file:

```
const MongoDB = require('../models/mongo')
```

7. Now, let's get a little fancier with our test suite. Augment the suite with the following code *inside* our `describe` method:

```
let db

beforeAll(async () => {
    db = await MongoDB.connectDB('test')
})

afterAll(async (done) => {
    await db.collection('names').deleteMany({})
    await MongoDB.disconnectDB()
    done()
})
```

And add the following case after our simple test:

```
it('should find names and return true', async () => {
    const names = await db.collection("names").find().toArray()
    expect(names.length).toBeGreaterThan(0)
})
```

Then run it with the same command as before: `node_modules/.bin/jest test.test.js`.

What's happening here? First of all, before every individual test in our test suite, we're specifying to connect to the database as per the method we wrote in our MongoDB model. After everything is done, tear down the database and disconnect.

And what happens when we run it? An epic failure!

```
$ node_modules/.bin/jest test.test.js
  FAIL  tests/test.test.js
    test
      ✓ should return true (2ms)
      X should find names and return true (9ms)

    ● test > should find names and return true

      expect(received).toBeGreaterThan(expected)
```

```
Expected: > 0
Received:   0

    20 |   it('should find names and return true', async () => {
    21 |     const names = await db.collection("names"
                ).find().toArray()
  > 22 |     expect(names.length).toBeGreaterThan(0)
       |                                          ^
    23 |   })
    24 | });

    at Object.<anonymous> (tests/test.test.js:22:26)

Test Suites: 1 failed, 1 total
Tests:       1 failed, 1 passed, 2 total
Snapshots:   0 total
Time:        1.622s, estimated 2s
Ran all test suites matching /test.test.js/i.
```

We should be expecting an error because we haven't yet *inserted* any information into a collection named `names` (or any other data!). Welcome to TDD: we wrote a test that fails before we wrote the code to make it pass.

Obviously, our next step in the process is to actually insert some data! Let's do that.

Storing and retrieving data

Let's work with a test suite that I wrote to help make sure that our MongoDB connection is a bit more robust and includes inserting data into the database and then testing to be sure it exists:

1. Examine `test/setup.model.test.js`:

```
const MongoDB = require('../models/mongo')
const insertRandomNames = require('../models/setup')

describe('insert', () => {
 let db

 beforeAll(async () => {
   db = await MongoDB.connectDB('test')
 })

 afterAll(async (done) => {
   await db.collection('names').deleteMany({})
```

```
      await MongoDB.disconnectDB()
      done()
   })

   it('should insert the random names', async () => {
     await insertRandomNames()

     const names = await db.collection("names").find().toArray()
     expect(names.length).toBeGreaterThan(0)
   })

})
```

2. If we run `node_modules/.bin/jest setup`, we'll see success because the
   `insertRandomNames()` method exists from our setup model. So let's take a look
   at our setup model (`models/setups.js`) and see what it's doing to populate the
   database:

```
const fs = require('fs')
const MongoDB = require('./mongo')

let db

const setup = async () => {
 db = await MongoDB.connectDB()
}

const insertRandomNames = async () => {
 await setup()

 const names = JSON.parse(fs.readFileSync(`${__dirname}/../
  data/starship-names.json`)).names

 const result = await db.collection("names").updateOne({ key:
  "names" }, { $set: { names: names } }, { upsert: true })

 return result
}

module.exports = insertRandomNames
```

3. Not too bad! We have one exported method that inserts names into the database based on a JSON file of "random" starship names that I've provided. The file is read and then put into the database as follows:

```
db.collection("names").updateOne({ key: "names" }, { $set: { names:
names } }, { upsert: true })
```

Since we're not getting too far into the guts of MongoDB itself, suffice to say that this line translates to "in the `names` collection (even if it doesn't yet exist), set the `names` key to equal the JSON. Update or insert as necessary".

We can now populate our database with ship names that we'll use from here on out. Execute `npm run install-data`.

So far, so good! There are many files in this project, so we won't walk through *all* of them; let's examine a representational sample.

Models, views, and controllers

The **Model-View-Controller** (**MVC**) paradigm is what we're using here within Express. While not really necessary in Express, I find the logical separation of concerns is useful and easier to work with than monolithic types of undifferentiated files. Before we go too far, I will mention that MVC could be considered an outdated pattern, as it does create some additional dependencies between layers. With that being said, the ideas behind an architectural paradigm that separates logic into discrete actors are sound in MVC. You may hear **MV*** used instead, which basically should be read as "model, view, and whatever that binds them together." MV* is more popularly used these days in certain frameworks.

The MVC construction separates the logic of the program into three parts:

1. **Models** deal with data interaction.
2. **Views** handle the presentation layer.
3. **Controllers** handle data manipulation and serve as the glue between the models and the views.

Here's a visual representation of the design pattern:

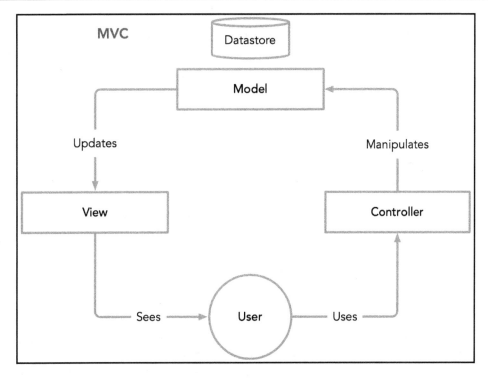

Figure 18.4 – The MVC paradigm's lifecycle

One of the more important parts to understand about this separation of concerns is that the **View** layer and the **Controller** layer are *never* supposed to interact directly with the **Datastore**; that honor is reserved for the models.

Now let's look at a view:

views/index.hbs

```
<h1>Starship Fleet</h1>

<hr />

<h2>Fleet Status</h2>
{{#if ships.length}}
 <table class="table">
   <tr>
     <th>Name</th>
     <th>Registry</th>
     <th>Top Speed</th>
     <th>Shield Strength</th>
```

```
        <th>Phaser Power</th>
        <th>Hull Damage</th>
        <th>Torpedo Complement</th>
        <th></th>
      </tr>
    {{#each ships}}
      <tr data-ship="{{this.registry}}">
        <td>{{this.name}}</td>
        <td>{{this.registry}}</td>
        <td>{{this.speed}}</td>
        <td>{{this.shields}}</td>
        <td>{{this.phasers}}</td>
        <td>{{this.hull}}</td>
        <td>{{this.torpedoes}}</td>
        <td><a class="btn btn-primary scuttle">Scuttle Ship</a></td>
      </tr>
    {{/each}}
    </table>
  {{else}}
    <p>The fleet is empty. Create some ships below.</p>
  {{/if}}
```

Express is controlling our views, and we're using Handlebars to handle our templating logic and loops. While simple in syntax, Handlebars is powerful and can greatly simplify our lives. In this case, we're testing for and looping over the ships variable in order to create a table of the ships we have or to send a message that the fleet is empty. How does our view get ships? It's given to the view from our **controller** by way of our **route**. Here's what that looks like for this portion:

routes/index.js

```
var express = require('express');
var router = express.Router();
const ShipsController = require('../controllers/ships');

/* GET home page. */
router.get('/', async (req, res, next) => {
 res.render('index', { ships: await ShipsController.getFleet() });
});

module.exports = router;
```

Why are we using `var` here instead of `const` or `let`? And why the semicolons? The answer is this: at the time of writing, the Express scaffolding tool still uses `var` and semicolons. It's always a best practice to standardize, but in this example I wanted to call attention to this fact. Feel free to standardize on the newer syntax as we work forward.

And now for the `getFleet` method:

controllers/ships.js

```
exports.getFleet = async (enemy = false) => {
  return await ShipsModel.getFleet(enemy)
}
```

Because this is a simple example, our controller doesn't do much besides getting information from the model, which queries MongoDB. Let's take a look at that:

models/ships.js

```
exports.getFleet = async (enemy) => {
  await setup()

  const fleet = await db.collection((!enemy) ? "fleet" :
  "enemy").find().toArray();
  return fleet.sort((a, b) => (a.name > b.name) ? 1 : -1)
}
```

The setup function dictates the connection to MongoDB (note the async/await setup!) and our fleet is either from the enemy or our fleet collection. The `return` line contains a convenience to sort the fleet by alphabetical order.

In this example, we're going to keep the controllers fairly simple and rely on the model to do the heavy lifting. This is a stylistic decision, though it's good to pick one side of the application to do the majority of your work.

It's time to look at the program from end to end.

Wiring your API together

To further understand the gameplay, we'll walk through the steps involved in firing a torpedo from a ship:

1. Locate the frontend JavaScript in `public/javascripts/play.js`:

```
document.querySelectorAll('.fire').forEach((el) => {
  el.addEventListener('click', (e) => {
    const weapon = (e.target.classList.value.indexOf('fire-torpedo')
    > 0) ? "torpedo" : "phasers"
    const target = e.target.parentNode.getElementsByTagName
    ('select')[0].value
```

2. Here we've made a click handler on the `fire` buttons in our interface and identified our weapon and target ship:

```
fetch(
`/play/fire?
attacker=${e.target.closest('td').dataset.attacker}&target=${target
}&weapon=${weapon}`)
.then(response => response.json())
.then(data => {
```

This line might take a bit of unpacking. We're making an AJAX call to our Node application from our JavaScript with certain query string parameters: `attacker`, `target`, and `weapon`. We're also expecting JSON to be returned from our application.

3. Remember that our backticks allow us to *compose* a string with variables in `${ }`:

```
const { registry, name, shields, torpedoes, hull, scuttled } =
data.target
```

4. We're using **object destructuring** to extract each piece of the information contained within `data.target`. This is a bit more efficient than defining them one by one or even with a loop, right?

```
if (scuttled) {
    document.querySelector(`[data-ship=${registry}]`).remove()
    document.querySelectorAll(`option[value=${registry}]`).
     forEach(el => el.remove())

    const titleNode = document.querySelector("#modal-1-title")

    if (data.fleet.length === 0) {
```

```
                titleNode.innerHTML = "Your fleet has been destroyed!"
            } else if (data.enemyFleet.length === 0) {
                titleNode.innerHTML = "You've destroyed the Borg!"
            } else {
                titleNode.innerHTML = `${name} destroyed!`
            }
            MicroModal.show('modal-1')
            return
        }
```

5. Our target ship has been destroyed if `scuttled` is `true`, so let's communicate that to the user. We're going to edit the values of our ship in either case:

```
            const targetShip = document.querySelector(`[data-
            ship=${registry}]`)

            targetShip.querySelector('.shields').innerHTML = shields
            targetShip.querySelector('.torpedoes').innerHTML = torpedoes
            targetShip.querySelector('.hull').innerHTML = hull
        })
    })
})
```

So that's the frontend code. If we look at our `app.js` file, we can see that our AJAX call to `/play` goes to the `playRouter` from an `app.use` statement. Therefore, our next stop is the router:

routes/play.js

```
const express = require('express');
const router = express.Router();
const ShipsController = require('../controllers/ships');

router.get('/', async (req, res, next) => {
 res.render('play', { fleet: await ShipsController.getFleet(), enemyFleet:
  await ShipsController.getFleet(true) });
});

router.get('/fire', async (req, res, next) => {
 res.json(await ShipsController.fire(req.query.attacker, req.query.target,
  req.query.weapon));
});

module.exports = router;
```

Since our URL was constructed from /play/fire, we know that the second router.get statement is the one handling our request. Onward to the controller and its fire method:

controllers/ships.js

```
exports.fire = async (ship1, ship2, weapon) => {
 let target = await ShipsModel.getShip(ship2)
 const source = await ShipsModel.getShip(ship1)
 let damage = calculateDamage(source, target, weapon)
  target = await ShipsModel.registerDamage(target, damage)

 return { target: target, fleet: await this.getFleet(false), enemyFleet:
  await this.getFleet(true) }
}
```

In the preceding code, we see the glue between the controller and the model. First of all, we're getting the target and source ships. Why do you think I decided to use let for the target and const for the source? If you reasoned that the target will need to be mutable, you're right: when we use the registerDamage method on our target, it'll be more efficient to rewrite the variable instead of creating a new one.

Before we look at our model's registerDamage method, acknowledge that the return path so far is that the controller will return to the route that returns to our frontend script.

Onward!

models/ships.js

```
exports.registerDamage = async (ship, damage) => {
 const enemy = (!ship.registry.indexOf('NCC')) ? "fleet" : "enemy"
  const target = await db.collection(enemy).findOne({ registry:
   ship.registry })

 if (target.shields > damage) {
   target.shields -= damage
 } else {
   target.shields -= damage
   target.hull += Math.abs(target.shields)
   target.shields = 0
 }

  await db.collection(enemy).updateOne({ registry: ship.registry }, { $set:
{ shields: target.shields, hull: target.hull } })
  if (target.hull >= 100) {
   await this.scuttle(target.registry)
   target.scuttled = true
 }
```

```
  return target
}
```

Now *here* is where we're actually communicating with our database. We can see that we're retrieving our target, registering damage to its shields and possibly its hull, setting those values in MongoDB, and eventually returning the target ship's information back through the controller to eventually arrive at our frontend JavaScript.

Let's take a look at this line:

```
await db.collection(enemy).updateOne({ registry: ship.registry }, { $set: {
shields: target.shields, hull: target.hull } })
```

We're going to update one item in the collection to state whether it's an enemy ship or in our fleet, and set the shield strength and hull damage.

Exporting functions

By now you may have noticed that some of the model methods, such as `registerDamage`, are prefaced with `exports` while others, such as `eliminateExistingShips`, are not. One aspect of good design in complex JavaScript applications is encapsulating the functions that are not designed to be used outside of a certain context. When prefaced with `exports`, a function can be invoked from a different context, such as from our controller. If it's not designed to be exposed to the rest of the application; in essence, it's a private function. The concept of exporting a variable is similar to the concept of scope in that we're making sure to keep our application clean and expose only the useful bits of the program.

If we take a look at `eliminateExistingShips`, we can see it's just a helper function used by `createRandom` to make sure we're not assigning the same ship registry number or name to two different ships. We can see this usage here in `createRandom`:

```
const randomSeed = Math.ceil(Math.random() * names.length);

const shipData = {
  name: (!enemy) ? names[randomSeed] : "Borg Cube",
```

more code... then:

```
while (unavailableRegistries.includes(shipData.registry)) {
  shipData.registry = `NCC-${Math.round(Math.random() * 10000)}`;
}
```

To make sure that our ship's registry number is unique to our fleet, we'll use a `while` loop to keep updating the ship's registry number until it's not one that already exists. Using the `eliminateExistingShips` helper function, we return and destructure the names and registries that already exist in our fleet so that we do not create duplicate registries.

We haven't used `while` loops often, as they are often blocking points in a program and can be easily misused. With that being said, this is a good use case for a `while` loop: it ensures that our program cannot continue unless the ship's registry is unique. With a randomization multiplier of 10,000, it's unlikely that a duplicate random registry will be generated twice in a row, or more, so a `while` loop is appropriate.

So, to export or not to export, that is the question. The answer depends on whether we need to use the function outside of its immediate scope. If there's no use for the function in another part of the program, then we shouldn't export it. In this case, our need to identify whether a ship's details already exist in the fleet is really only useful in our `ships` model, so we'll refrain from exporting it.

Improving our program

As you read through the `ships` model and controller, I'm sure you can find areas for improvement. For example, the way I've written the switches for understanding whether the ship was in our fleet or the enemy fleet is a bit rigid: it wouldn't be able to accommodate three separate fleets in one battle. Every programmer creates **technical debt**, or small errors or inefficiencies in code. This then necessitates **refactoring**, the practice of altering code to make it better. Don't be fooled into thinking you've ever written *The Perfect Program*—such a thing does not exist. Improvement and continual iteration are part of the programming process.

There is a major caveat to refactoring, however, and that's something often called a **contract**. When designing a backend designed to be used by the frontend, and when different parties are writing different parts of a system, it's important to be in sync with each other and the premise and needs of the program as a whole.

Let's take, for example, our frontend JavaScript code. If we enumerate the endpoints it's using, we'll see four endpoints being used:

- `/ships`
- `/ships/${e.currentTarget.closest('tr').dataset.ship}`

- `/ships/random`
- `/play/fire?attacker=${e.target.closest('td').dataset.attacker}&`
  `target=${target}&weapon=${weapon}`` `

At a minimum, when we refactor our backend code, we should assume a contractual obligation to change neither the path of these endpoints nor the expectation of datatype(s) to be received.

One way we can help make our code more future-proof is *inline documentation* using a loose standard called JSDoc. The creation of documentation from code comments is a long-standing practice and comment structures exist for many languages in order to facilitate a standard. In cases such as APIs, often a helper program is run against the source code to generate standalone documentation, often as a small HTML/CSS microsite. You may have encountered unrelated programs with similarly styled online documentation. There's a good possibility those unrelated documentation sites were generated from code by the same mechanism.

Why is this important in a chapter about MongoDB? Well, documentation isn't only a need for database usage; rather, it is important when creating any type of program that has multiple moving parts. Consider the last endpoint in the preceding list:
`/play/fire?attacker=${e.target.closest('td').dataset.attacker}&target=$`
`{target}&weapon=${weapon}``.

The fire endpoint takes three parameters: `attacker`, `target`, and `weapon`. But what are those parameters? What do they look like—are they objects? Strings? Booleans? Arrays? Additionally, if we're going to accept user-generated data, we need to be a bit more careful than we have been, because of **GIGO: Garbage In, Garbage Out**. If we populate our database with bad data, the best we can expect is a broken program. In fact, the worst we can expect is a security **compromise**: the leaking of database or server credentials or malicious code execution. Let's talk about security.

Security

If you're familiar with SQL, you may be familiar with a security vulnerability known as **SQL Injection**. Good information on web application security best practices can be found at `owasp.org`. The **Open Web Application Security Project (OWASP)** is a community-driven initiative to catalog and educate users on the security vulnerabilities present in web applications in order that we can more effectively combat against malicious hackers. If you've ever had your email, social account, or website hacked, you know the pain that ensues—digital identity theft. OWASP's listing for SQL injection is here: `https://owasp.org/www-community/attacks/SQL_Injection`.

So why are we talking about SQL if we're using a NoSQL database in the form of MongoDB? Because *SQL injection doesn't exist in MongoDB*. "GREAT!", you might say, "My security woes are solved!" This, unfortunately, is not the case. Coupled with the idea of refactoring to improve efficiency of an application, refactoring to mitigate security intrusion vectors is an important part of being responsible for a web application. I've worked at a company that was hacked—and it was because of less than five characters placed into a URL. This enabled a hacker to break the operation of the web application and perform arbitrary SQL commands. Sanitizing and refactoring for security all user-generated content is an essential part of web security. Now, we haven't done that for this application because I trust you're not going to hack your own machine.

Wait. Didn't I just say that SQL injection doesn't exist with MongoDB? Yes, and NoSQL databases have their equivalent method of attack: **code and command injection**. Because we haven't sanitized, or verified the integrity of, our user input, it's possible for our application to store and use arbitrary code that's been submitted and stored in our database. While a full primer on JavaScript security is not within scope of this book, do keep it in mind. The long story short is to sanitize, or verify the validity of, your user-generated input.

And with that, let's wrap up this chapter. Just remember to stay safe when writing MongoDB applications in the wild!

Summary

JavaScript doesn't exist in isolation! MongoDB is a great companion to JavaScript as it is designed to be object-oriented and relies on a JavaScript-friendly querying syntax. We've learned the principles behind TDD, worked with the MVC paradigm, and expanded our game a bit more.

As with all coding exercises, be sure to consider the use cases when using a database such as MongoDB: while MongoDB's syntax isn't vulnerable to SQL injections, it is still vulnerable to other types of injections that can compromise your application.

Hopefully, our starship game is interesting enough for you to keep developing it. Our next (and final) chapter wraps together our principles of JavaScript development and polishes our game.

19
Putting It All Together

At last! We can now build both the front- and backend of a website and use JavaScript on both sides! To bring it all together, let's build a small web application that uses an Express API with a React frontend and MongoDB.

For our final project, we'll use our skills to create a database-backed travelogue or travel journal, complete with photos and stories. Our approach will be to take this from an initial visual layout, all the way through to the front- and backend code. If your HTML/CSS skills aren't great, don't worry: the code is provided for you at multiple instances so you can begin working on your project where you'd like.

The following topics will be covered in this chapter:

- The project brief
- Scaffolding – React
- The backend – setting up our API
- The database – all CRUD operations

Technical requirements

Be prepared to work with the code provided in the chapter-19 directory of the repository, at https://github.com/PacktPublishing/Hands-on-JavaScript-for-Python-Developers/tree/master/chapter-19. As we'll be working with command-line tools, also have your Terminal or command-line shell available. We'll need a modern browser and a local code editor.

The project brief

When beginning a real-world web project from start to finish, it's important to gather the **requirements** up front. This can be presented in many forms: a verbal description, a bulleted list of features, a visual wireframe, a complete design document, or any combination of these. When examining requirements, it's important to be as explicit as possible in order to minimize miscommunication, redundant or abandoned work, and a streamlined workflow. For this project, we'll begin with visual comps.

If you've ever worked with a graphic designer, you're probably familiar with the term comp. A visual comp, short for *comprehensive layout*, is a design artifact that is a high-fidelity visual representation of the desired end state for a project. For example, a print project's comp would be a digital file with all the required assets ready to send to the printer for immediate use. For digital work, you may receive Adobe Photoshop, XD, or Sketch files, or many other types of design document formats.

Let's take a look at the visuals so that we can then ascertain our requirements:

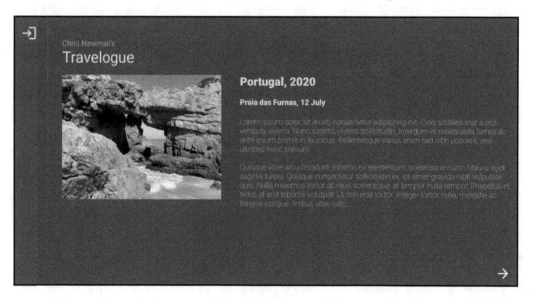

Figure 19.1 – Main page

Our application will have *logged in* and *logged out* states. When logged out, the user will see a cover page and will be able to explore the travelogue's entries with navigation buttons. For a challenge, display a random entry on page load.

The login button in the top-left corner will lead to the next screen, the **Log In** screen:

Figure 19.2 – Log In

The login page can be as simple or as complex as you'd like. Perhaps entering any username and password combination will work, or for an extra challenge, you could incorporate Google or Facebook authentication. You could even write your own authentication using your database to store credentials.

Once authenticated, we have a new button in the left bar: the dashboard button. This is what takes us to the various parts of the application:

Figure 19.3 – Dashboard

When the **Countries Visited** button is clicked, we'll display this vector map powered by the D3.js graphics library:

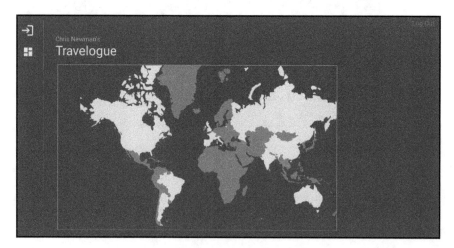

Figure 19.4 – Travel map

The countries highlighted are controlled via a JSON manifest, fed by the database of entries.

And last but not least, the user needs to be able to write entries and insert photos:

Figure 19.5 – New entry/edit entry screen

We'll be using a JavaScript WYSIWYG (What you see is what you get) editor called Quill.

As you build the app, feel free to take some liberties with the look and feel of it—customize it to make it your own! You might also want to add a few more features, such as a media library to manage uploaded photos, or a search function.

Now that we have an idea of our visual layout, let's get started on the frontend of our project.

Scaffolding – React

Our project is a prime candidate for using React for our frontend, so let's outline our requirements for the frontend: *One single React application with reusable components and Hooks and context for state saving*. Hooks are a new concept compared to our previous use of React. Added in React 16.8, Hooks are functions that allow you to manipulate the state and context for state management within a functional component.

In addition to our handcrafted React app, we'll incorporate a few additional pre-built libraries to simplify our project and to utilize ready-made tools. D3.js is a powerful graphics and data visualization library that we'll leverage for our map. Quill is a rich text editor that will allow you to write the entries with text formatting, and upload and place photos.

It's up to you to decide whether you'd like to start with `npx create-react-app` or use the scaffolded code provided in the `Step 1` folder in the `chapter-19` directory of the GitHub repository.

I'm going to make a few recommendations on additional packages to use; as you go through the project, feel free to add or subtract packages. I'll be using the following:

- Bootstrap (for layout)
- `d3`, `d3-queue`, and `topojson-client` (for our map)
- `node-sass` (for more efficient stylesheets using Sass)
- `quill` and `react-quilljs` (a WYSIWYG editor)
- `react-router-dom` (a React extension for URL pathing made easy)
- `react-cookie` (a package to easily use cookies)

If you're starting from scratch, feel free to get set up with the `create-react-app` scaffold now, or begin using the `Step 1` directory. For the rest of this chapter, instructions are provided for you to follow along step by step.

Inside the `Step 1` directory, you'll find the following:

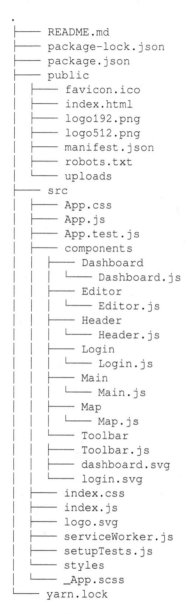

```
.
├── README.md
├── package-lock.json
├── package.json
├── public
│   ├── favicon.ico
│   ├── index.html
│   ├── logo192.png
│   ├── logo512.png
│   ├── manifest.json
│   ├── robots.txt
│   └── uploads
├── src
│   ├── App.css
│   ├── App.js
│   ├── App.test.js
│   ├── components
│   │   ├── Dashboard
│   │   │   └── Dashboard.js
│   │   ├── Editor
│   │   │   └── Editor.js
│   │   ├── Header
│   │   │   └── Header.js
│   │   ├── Login
│   │   │   └── Login.js
│   │   ├── Main
│   │   │   └── Main.js
│   │   ├── Map
│   │   │   └── Map.js
│   │   ├── Toolbar
│   │   │   Toolbar.js
│   │   ├── dashboard.svg
│   │   └── login.svg
│   ├── index.css
│   ├── index.js
│   ├── logo.svg
│   ├── serviceWorker.js
│   ├── setupTests.js
│   └── styles
│       └── _App.scss
└── yarn.lock
```

It's a standard `create-react-app` scaffold, with a few differences from what we've done before. Let's examine one component: the header.

Functional components

Here's the code for our `Header.js` file:

```
import React from 'react'

function Header() {
 return (
   <>
     <h2>Chris Newman's</h2>
     <h1>Travelogue</h1>
   </>
 )
}
export default Header
```

You should notice a few things: first of all, the filename ends in `js`, *not* `jsx`. Next, our component is a function that returns HTML, versus a class extending `React.Component`. While both class-based and functional components are valid in React, functional components are considered more modern when using React, especially with the latest methods to leverage state and context. We won't get into the difference between functional and object-oriented programming right now, but suffice to say there are differences to be aware of. You can find a useful resource on these differences at the end of the chapter.

Next steps

To progress the app to the next stage, consider the functional requirements we laid out. A good next step may be to implement a login system. At this point, you probably would neither want nor need to actually validate credentials, so a dummy login page will suffice. You have the markup in `Login/Login.js`.

The approach we're going to take is to use **Hooks** and **context**. Since this is a fairly involved topic, we won't get into all the details here, but there are plenty of articles explaining the concepts. Here's one: `https://www.digitalocean.com/community/tutorials/react-crud-context-hooks`.

We'll go through one example of context and a couple of examples of Hooks to get you started:

1. First, we need to create a `UserContext.js` file, which will help track our login state throughout the life cycle of our user interaction. The code itself is quite simple:

   ```
   import React from 'react'

   export const loggedIn = false

   const UserContext = React.createContext(loggedIn)

   export default UserContext
   ```

2. React's **Context API** is a method of providing stateful information to multiple components. Notice how I said "provide"? That's exactly what we need to do next: provide our `App.js` context. We wrap the component as such:

   ```
   import React, { useState } from 'react'
   import './styles/_App.scss'
   import Main from './components/Main/Main';
   import UserContext, { loggedIn } from './components/UserContext'

   function App() {

     const loginHook = useState(loggedIn)

     return (
       <UserContext.Provider value={loginHook}>
         <div className="App">
           <Main />
         </div>
       </UserContext.Provider>
     )
   }

   export default App
   ```

   Notice how we've imported `UserContext` and wrapped our `App` component in the `UserContext.Provider` tags with the `loginHook` stateful value provided to it, and thereby to its children.

3. Our `Main.js` file needs some changes too. Take a look at this snippet:

   ```
   function Main() {
     const [loggedIn, setLoggedIn] = useContext(UserContext)
   ```

```
const [cookies, setCookie] = useCookies(['logged-in'])
...
```

We need to import `useContext` and `useCookies` from React and `react-cookies`, respectively, and then we can use these **Hooks** to work with our login state. In addition to the internal context, we're also going to store our login status within a cookie to allow returning sessions to remain logged in. We also want to import `useEffect` from React for our next step:

```
const setOrCheckLoggedIn = (status) => {
    if (cookies['logged-in'] === 'true' || status) {
      setLoggedIn(true)
    }

    if (status && cookies['logged-in'] !== 'true') {
      setCookie('logged-in', true)
    }
}

useEffect(() => {
setOrCheckLoggedIn()
})
```

Do you remember how, in previous chapters, we reacted to the mount states of React components directly with `componentDidMount()`? With React Hooks, we can use the `useEffect` Hook to work with the state of our component. Here, we're going to ensure that our user context (`loggedIn`) and the `logged-in` cookie are set appropriately.

4. Our `setOrCheckLoggedIn` function also needs to be passed to other components, namely `Toolbar` and `Login`. Set it as the `doLogin` prop.

From this point forward, when we include the context of `UserContext`, we can rely on the `loggedIn` state variable to determine whether or not our user is logged in. For example, our simple `Login` component's logic could utilize these Hooks as follows:

```
import React, { useContext } from 'react'
import UserContext from '../UserContext'

const Login = (props) => {

let [loggedIn, setLoggedIn] = useContext(UserContext)

  const logMeIn = () => {
    loggedIn = !loggedIn
```

```
      props.doLogin(loggedIn)
    }

    return (
      <>
        <div className="Login">
          <h1>Log In</h1>

          <p><input type="text" name="username" id="username" /></p>
          <p><input type="password" name="password" id="password"
          /></p>
          <p><button type="submit" onClick={logMeIn}>Go</button></p>
        </div>
      </>
    )
  }

export default Login
```

Fairly straightforward! First, we get our context and upon clicking the Go button, we flip the context. You should incorporate similar logic in the `Toolbar.js` file for the login icon to also handle logging out.

Now, we're going to need a backend to interact with our frontend and broker the transactions with the MongoDB database, which will store our story entries and possibly our user authentication data. It will also be necessary to create an endpoint to upload images, as frontend code alone *cannot* write to a server's filesystem.

The backend – setting up our API

Let's catalog the endpoints we'll need to make our travelogue work:

- *Read (GET):* Like most APIs, we'll need an endpoint to read entries. We won't force authentication or being logged in for this.
- *Write (POST):* This endpoint will be used for both creating a new trip and editing existing ones.
- *Upload (POST):* We'll need an endpoint to call from our frontend in order to upload photos.
- *Login (POST) (Optional):* If you'd like to take your authentication into your own hands, create a login endpoint that can use credentials from the database or a social-media login endpoint.

- *Media (GET) (Optional):* It will be useful to have an API that lists all of the media files uploaded to your server.
- *Countries (GET) (Optional):* It will also be a good idea to have an endpoint specifically for listing the countries you've visited to power your world map.

You may find yourself creating more endpoints as you work, and that's fine! It's always a good idea to plan your API from start to finish, but if you need to make changes along the way to make your life easier with helper endpoints or other parts, feel free.

We're ready to move on to the Step 3 directory in our repository.

API as a proxy – Step 3

Because we're using a React frontend, we'll revisit the idea of using Express as a backend with React proxying our API requests, as follows:

1. The first thing we need to do is to tell our system to use a proxy by adding this line to our package.json: "proxy": "http://localhost:5000".

2. After adding that, restart React (you'll notice that our frontend homepage has changed; we'll get to that in a moment) and then, in the api directory, execute npm install in the api directory and then npm start.

3. We should test our backend to be sure our API is responding. Add this as a test to the App.js file after the imports:

```
fetch('/api')
  .then(res => res.text())
  .then(text => console.log(text))
```

This very basic fetch call should call the routes/index.js component's get method in our API:

```
router.get('/', (req, res) => {
 res.sendStatus(200)
})
```

At this point, our console should display OK. If you have any problems at this stage, it would be advisable to debug them now.

4. We know we'll be setting up a database to handle our data, but for the time being, we can scaffold our API's methods, as you can see in routes/index.js:

```
router.get('/article', (req, res) => {
 res.send({ story: "A story from the database" })
```

```
})

router.post('/article/edit', (req, res) => {
  res.sendStatus(200)
})

router.post('/media/upload', (req, res) => {
  res.sendStatus(200)
})

router.get('/media', (req, res) => {
  res.send({ media: "A list of media" })
})

router.post('/login', (req, res) => {
  res.sendStatus(200)
})

router.get('/countries', (req, res) => {
  res.send({ countries: "A list of countries" })
})
```

Now that we've made the scaffold of our login system in **Step 2**, I've made a few alterations to the Step 3 directory. As mentioned before, our homepage is a little different in that it's the index page of the travelogue, used to display a story while the user is logged out.

5. Examine the Story/Story.js component next:

```
import React from 'react'

function Story() {

  fetch('/api/article')
    .then(res => res.json())
    .then(json => console.log(json))

  return (
    <div className="Story">
      <h1>Headline</h1>
    ...
```

Yes, another dummy API call to our backend! This call is also a simple GET request, so let's do something a bit more involved.

6. Go ahead and log in to the site and you'll see something different on your dashboard:

Figure 19.6 – Our dashboard is taking shape...

7. Great, now we have a full dashboard. Click the **ADD TRIP** button and you'll be presented with an editor, as follows:

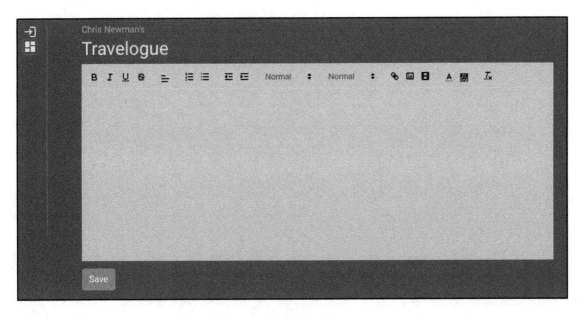

Figure 19.7 – Our text editor

If you enter rich text in the editor and save it, you'll see a response in your console from the API with the submitted data. From there, we need to work with our API to save the data to our database. So... last, but not least, we'll need to set up our database.

The database – all CRUD operations

We will, of course, need a datastore for our create, read, update, and delete functionality, so let's return to MongoDB in order to store these documents. You can refer back to Chapter 18, *Node.js and MongoDB,* if you need to refresh your memory on getting set up.

To get started with setting up a database from scratch, it helps to think of the database structure that you intend to use. While MongoDB doesn't require a schema, it's still a good idea to plan your MongoDB documents so that you're not being arbitrary with functionality or naming between sections.

Here's an idea of what each collection might look like:

```
settings:  {
  user
    firstname
    lastname
    username
    password
  title
  URL
  media directory
}

entry: {
  title
  location
  date
    month
    day
    year
  body
}

location: {
  city
  region
  country
  latitude
  longitude
  entries
}
```

It's good to keep your database simple, but remember that you can always expand on it.

Summary

Now, of course, I couldn't just *hand* you a final project, could I? In this chapter, we scaffolded our travelogue—the rest is up to you. There are a few pieces yet to do to have a fully functional project. After all, we haven't fully adhered to our visual comps, have we? Here are some ideas of what kinds of functionalities to implement in order to complete the project:

- Persist information to the database.
- Work on image uploading and saving.
- Edit existing articles.
- Create the `countries` endpoint to populate the D3.js map.
- Enable true login.
- Streamline the user journey.

When complete, this project will be a piece for your portfolio to show how *you*, a Python developer, mastered JavaScript. From the beginnings of datatypes, grammar, loops, and Node.js, to finally creating a fully functional project, you've come a long way.

It's with gratitude that I thank you for accompanying me on this journey! Keep learning, and **live long and prosper**.

Further reading

A useful resource on the differences between functional programming and object-oriented programming can be found at `https://www.geeksforgeeks.org/difference-between-functional-programming-and-object-oriented-programming/`.

Assessments

Chapter 1

1. What international group maintains the official specification for JavaScript?
    1. W3C
    2. **Ecma International**
    3. Netscape
    4. Sun
2. Which backends can communicate with JavaScript?
    1. PHP
    2. Python
    3. Java
    4. **All of the above**
3. Who was the original author of JavaScript?
    1. Tim Berners-Lee
    2. **Brendan Eich**
    3. Linus Torvalds
    4. Bill Gates
4. What is the DOM?
    1. JavaScript's representation of HTML in memory
    2. An API to allow JavaScript to modify the page
    3. **Both of the above**
    4. None of the above
5. What is the primary use of Ajax?
    1. Communication with the DOM
    2. Manipulation of the DOM
    3. Listening for user input
    4. **Communication with a backend**

Chapter 2

1. **True** or false: Node.js is single-threaded.
2. True or **false**: The architecture of Node.js makes it impervious to **Distributed Denial of Service (DDoS)** attacks.
3. Who originally created Node.js?
    1. Brendan Eich
    2. Linux Torvalds
    3. Ada Lovelace
    4. **Ryan Dahl**
4. True or **false**: JavaScript on the server side is inherently insecure because the code is exposed on the frontend.
5. True or **false**: Node.js is inherently superior to Python.

Chapter 3

1. Which of these is not a valid JavaScript variable declaration?
    1. var myVar = 'hello';
    2. const myVar = "hello"
    3. **String myVar = "hello";**
    4. let myVar = "hello"
2. Which of these starts a function declaration?
    1. **function**
    2. const
    3. func
    4. def
3. Which of these is not a basic loop type?
    1. for..in
    2. for
    3. while
    4. **map**
4. JavaScript *requires* line delineation with semicolons:
    1. True
    2. **False**

5. Whitespace *never* counts in JavaScript:
    1. True
    2. **False**

Chapter 4

1. JavaScript is inherently:
    1. Synchronous
    2. Asynchronous
    3. **Both**

2. A `fetch()` call returns a:
    1. `then`
    2. `next`
    3. `finally`
    4. **Promise**

3. With prototypal inheritance, we can (select all options that apply):
    1. **Add methods to a base data type.**
    2. **Subtract methods from a base data type.**
    3. Rename our data type.
    4. Cast our data into another format.

```
let x = !!1
console.log(x)
```

4. Given the preceding code, what will be the expected output?
    1. 1
    2. false
    3. 0
    4. **true**

```
const Officer = function(name, rank, posting) {
 this.name = name
 this.rank = rank
 this.posting = posting
 this.sayHello = () => {
 console.log(this.name)
 }
}
```

```
         const Riker = new Officer("Will Riker", "Commander", "U.S.S.
         Enterprise")
```

5. Given this code, what's the best way to output "Will Riker"?
   1. `Riker.sayHello()` *
   2. `console.log(Riker.name)`
   3. `console.log(Riker.this.name)`
   4. `Officer.Riker.name()`

Chapter 5

Consider the following code:

```
function someFunc() {
  let bar = 1;

  function zip() {
    alert(bar); // 1
    let beep = 2;

    function foo() {
      alert(bar); // 1
      alert(beep); // 2
    }
  }

  return zip
}


function sayHello(name) {
  const sayAlert = function() {
    alert(greeting)
  }

  const sayZip = function() {
    someFunc.zip()
  }

  let greeting = `Hello ${name}`
  return sayAlert
}
```

1. How would you get an alert of `'Hello Bob'`?
   1. `sayHello()('Bob')`
   2. `sayHello('Bob')()` *
   3. `sayHello('Bob')`
   4. `someFunc()(sayHello('Bob'))`

2. What will `alert(greeting)` do in the preceding code?
   1. Alert `'greeting'`
   2. Alert `'Hello Alice'`
   3. **Throw an error**
   4. None of the above

3. How would we get an alert message of 1?
   1. `someFunc()()` *
   2. `sayHello().sayZip()`
   3. `alert(someFunc.bar)`
   4. `sayZip()`

4. How would we get an alert message of 2?
   1. `someFunc().foo()`
   2. `someFunc()().beep`
   3. **We can't, because it's not in scope**
   4. We can't, because it's not defined

5. How can we change `someFunc` to alert **1 1 2**?
   1. We can't.
   2. Add `return foo` after `return zip`.
   3. Change `return zip` to `return foo`.
   4. **Add** `return foo` **after the** `foo` **declaration**.

6. Given a correct solution to the preceding question, how would we actually get three alerts of **1, 1, 2**?
   1. `someFunc()()()` *
   2. `someFunc()().foo()`
   3. `someFunc.foo()`
   4. `alert(someFunc)`

Chapter 6

Consider the following code:

```
<button>Click me!</button>
```

1. What is the correct syntax to select the button?
    1. document.querySelector('Click me!')
    2. document.querySelector('.button')
    3. document.querySelector('#button')
    4. **document.querySelector('button')**

Take a look at this code:

```
<button>Click me!</button>
<button>Click me two!</button>
<button>Click me three!</button>
<button>Click me four!</button>
```

1. True or **False**: document.querySelector('button') will serve our needs to place a click handler on each button.
2. To change the text of the button from "Click me!" to "Click me first!", what should we use?
    1. **document.querySelectorAll('button')[0].innerHTML = "Click me first!"**
    2. document.querySelector('button')[0].innerHTML = "Click me first!"
    3. document.querySelector('button').innerHTML = "Click me first!"
    4. document.querySelectorAll('#button')[0].innerHTML = "Click me first!"
3. What method could we use to add another button?
    1. document.appendChild('button')
    2. document.appendChild('<button>')
    3. **document.appendChild(document.createElement('button'))**
    4. document.appendChild(document.querySelector('button'))
4. How can we change the class of the third button to "third"?
    1. document.querySelector('button')[3].className = 'third'
    2. **document.querySelectorAll('button')[2].className = 'third'**
    3. document.querySelector('button[2]').className = 'third'
    4. document.querySelectorAll('button')[3].className = 'third'

Chapter 7

Answer the following questions to gauge your understanding of events:

1. Which of these is the second phase of the event lifecycle?
    1. Capturing
    2. **Targeting**
    3. Bubbling

2. (Choose all the correct answers) What does the event object provide us with?
    1. **The type of event that is triggered**
    2. **The target DOM node, if applicable**
    3. **The mouse coordinates, if applicable**
    4. The parent DOM node, if applicable

Look at this code:

```
container.addEventListener('click', (e) => {
  if (e.target.className === 'box') {
    document.querySelector('#color').innerHTML =
e.target.style.backgroundColor
    document.querySelector('#message').innerHTML = e.target.innerHTML
    messageBox.style.visibility = 'visible'
    document.querySelector('#delete').addEventListener('click', (event) =>
{
      messageBox.style.visibility = 'hidden'
      e.target.remove()
    })
  }
})
```

1. Which JavaScript features is it using? Select all the answers that apply:
    1. **DOM manipulation**
    2. **Event delegation**
    3. **Event registration**
    4. **Style changes**

2. What will happen when the container is clicked?
    1. box will be visible.
    2. #color will be red.
    3. Both options 1 and 2.
    4. **There is not enough context to say.**

3. In which phase of the event lifecycle do we typically take action?
    1. **Targeting**
    2. Capturing
    3. Bubbling

Chapter 9

1. What is the root cause of memory problems?
    1. The variables in your program are global.
    2. **Inefficient code.**
    3. JavaScript's performance limitations.
    4. Hardware inadequacies.

2. When using DOM elements, you should store references to them locally versus always accessing the DOM.
    1. True
    2. False
    3. **True when using them more than once**

3. JavaScript is pre-processed on the server side, and thus more efficient than Python.
    1. True
    2. **False**

4. Setting breakpoints can't find memory leaks.
    1. True
    2. **False**

5. It's a good idea to store all variables in the global namespace as they're more efficient to reference.
    1. True
    2. **False**

Other Books You May Enjoy

If you enjoyed this book, you may be interested in these other books by Packt:

Svelte 3 Up and Running
Alessandro Segala

ISBN: 978-1-83921-362-5

- Understand why Svelte 3 is the go-to framework for building static web apps that offer great UX
- Explore the tool setup that makes it easier to build and debug Svelte apps
- Scaffold your web project and build apps using the Svelte framework
- Create Svelte components using the Svelte template syntax and its APIs
- Combine Svelte components to build apps that solve complex real-world problems
- Use Svelte's built-in animations and transitions for creating components
- Implement routing for client-side single-page applications (SPAs)
- Perform automated testing and deploy your Svelte apps, using CI/CD when applicable

Learning Angular - Third Edition

Aristeidis Bampakos, Pablo Deeleman

ISBN: 978-1-83921-066-2

- Use the Angular CLI to scaffold, build, and deploy a new Angular application
- Build components, the basic building blocks of an Angular application
- Discover techniques to make Angular components interact with each other
- Understand the different types of templates supported by Angular
- Create HTTP data services to access APIs and provide data to components
- Enhance your application's UX with Angular Material
- Apply best practices and coding conventions to your large-scale web development projects

Leave a review - let other readers know what you think

Please share your thoughts on this book with others by leaving a review on the site that you bought it from. If you purchased the book from Amazon, please leave us an honest review on this book's Amazon page. This is vital so that other potential readers can see and use your unbiased opinion to make purchasing decisions, we can understand what our customers think about our products, and our authors can see your feedback on the title that they have worked with Packt to create. It will only take a few minutes of your time, but is valuable to other potential customers, our authors, and Packt. Thank you!

Index